INFORMATION AS MATERIAL

DECIPHERING HUMAN CHROMOSOME 16:

INDEX TO THE REPORT

Sarah Jacobs,
co-ordinator

A copy of the report* can be downloaded free from the publisher www.informationasmaterial.com.

"Chromosome 16" by the Human Genome Project is published by
Project Gutenberg. Urbana Illionois (USA).
Etext #2216. First release June 2000.
This text is in the public domain.
The introduction contains the following information:
"The Project Gutenberg Etext of Human Genome Project, Chromosome 16
of 24 in the chromosome sequence....
"This was downloaded from ncbi.nlm.nih.gov/genome/seq
Just click on the chromosome 16 to get download options
"At the time of download approximately 11.7% of the
Human Genome had been mapped, about 378 Gigabytes worth.
NIH asked us to credit them this way, not as above:
this is still a rough draft, a year ahead of schedule
as the June, 2000 date is NOT an error, even though I am
writing this in June, 1999 ....
"**Accreditation:
"(http://www.ncbi.nlm.nih.gov/genemap98/ for GeneMap'98,
or http://www.ncbi.nlm.nih.gov/genemap/ for GeneMap'99)
as well as the GeneMap'98 paper:
"Deloukas, et al. 1998. A physical map of 30,000 human
genes. Science 282(5389), 744-746 (October 23)."

Cookie Lopez, design.

British Library Cataloguing in Publication Data.

A catalogue record for this book is available from the British Library.

INFORMATION AS MATERIAL

ISBN 978 0 9553092 2 9

*Deciphering Human Chromosome 16:
We Report Here

TAACCCTAACCCTAACCCTAACCCTAACCCTAACCGACCCTCACCCTCACCCTAACCACATGAGCAATGTGGGTGTTATATTTTAGCTGTCATGGGTGCATTAGG
AATGCTGCATTTGTGTTTCAACGCTGCAACTGGACCCTGCAATGCAGCCCCTCGCCTTGCCTTGGGAGAATCTCGGTGCCCAGGATTCAGAGGGGCTTTTAGTTT
CCCATTTTCCACACTGAACCGTTCTAACTGGTCTCTGACCTTGATTATTCACGGCTGCAACCGGGAAAGATTTTATTCACTGTCAATGCGCCCCGAGTTGTCCCAAA
GCCAGGCAGTGCCCCCAACGTCTGTGCTTAGCAGAATGCTGCTCCACCTTTACGGTGACCCCCAGGTCTGTGCTGAGCAGAACGCAGCTCCGCCCTCGCA
GTACCCTCAGCCCGCCCGCCCGGGTCTGACCTGAGCAGAACTCTGCTCTGCCTTCGCAGTACCACCGAAATCTGTGCAAAGGAGAACGCAGCTCCGCCC
TCGCGGTGCTCTCCGCGTCTGTGCTGAGGAGAACGCAACTCCGCCGTCGCAAAGGCGCGCGCCGCGCCGGCGCAGGCGCAGAGGGGCGCGCCGC
GCCGGCGCAGGCGCAGAGACACATGCTAGCGCGTCCAGGGGTGGAGGCGTGGCGCAGGCGCAGAGACGCACGCCTACGGGCGGGGGTTGGGGG
GGCGTGTGTTGCAGGAGCAAAGTCGCACGGCGCCGGCCTGGGGGCGGGGGGTGGGGGGCGCCGTTGCACGCGCAGAAACTCACGTCACGGTGGCG
CGGCGCAGAGACGGGTGGAACCTCAGTAATCCGAAAAGCCGGGATCGACCGCCCCTTGCTTGCAGCCGGGCACTACAGGACCCGCTTGCTCACGGTG
CTGTGCCAGGGCGCCCCCTGCTGGCGACTAGGGCAACTGCAGGGCTCTCTTGCTTAGAGTGGTGGCCAGCGCCCCCTGCTGGCGCCGGGGCACTGCA
GGGCCCTCTTGCTTACTGTATAGTGGTGGCACGCCGCCTGCTGGCAGCTAGGGACATTGCAGGCTCCTCTTGCTCAAAGTGTAGTGGCAGCACGCCCGCCT
GCTGGCAGCTGGGGACACTGCCGGGCCCTCTTGCTCCAACAGTAGTGGCGGATTATAGGGAAACACCCGGAGCATATGCTGTTTGGTCTCAGTAGACTCCT
AAATATGGGATTCCTGGGTTTAAAAGTATAAAATAAATATGTTTAATTTGTTAACTGATTACCATCAGAATTATACTGTTCTGTATCCCACCAGCAATGTCTAGGAATACCT
GTTTCTCCACAAAGTGTTTACTTTTGGATTTTTGCCAGTCTAGCAGGTGAAGCCCTGGAGATTCTTATTAGTCATTTGGGCTGGGGCCTGGCCATGTGTATTTTTTTAAATTT
CCACTGATGATTTTGCTGCATGGCCGGTGTTGAGAATGACTGCGCAAATTTGCCGGATTTCCTTTGCTGTTCCTGCATGTAGTTTAAACGAGATTGCCACCACCGG
GTATCATTCACCATTTTTCTTTTTGTTAACTTGCCGTCAGCCTTTTCTTTGACCTCTTCTTTCTCTTCATGTGTATTTGCTGTCTCTTAGCCCAGACTTCCCGTGTCCTTTCCACC
GGGCCTTTGAGAGGTCACAGGGTCTTGATGCTGTGGTCTTGATCTGCAGGTGTCTGACTTCCAGCAACTGCTGGCCTGTGCCAGGGTGCAAGCTGAGCACTGG
AGTGGAGTTTTCCTGTGGAGAGGAGCCATGCCTAGAGTGGGATGGGCCATTGTTCATCTTCTGGCCCCTGTTGTCTGCATGTAACTTAATACCACAACCAGGCAT
ACGGGAAAGATTGGAGGAAAGATGAGTGAGAGCATCAACTTCTCTGACAACCTAGGCCAGTAAGTAGTGCTTGTGCTCATCTCCTTGGCTGTGATACGTGGCC
GGCCCTCGCTCCAGCAGCTGGACCCCTACCTGCCGTCTGCTGCCATCGGAGCCCAAAGCCGAGCTGTGACTGCTCAGACCAGCCGGCTGGAGGGAGG
GGCTCAGCAGGTCTGGCTTTGGCCCTGGGAGAGCAGGTGGAAGATCAGGCAGGCCATCGCTGCCGCAGAACCCAGTGGATTGGCCTAGGTGGGATCTCT
GAGCTCAACAAGCCCTCTCTGGGTGGTAGGTGCAGAGAGGGGAGGGGCAGAGCCGCAGGCACAGCCAAGAGGGCTGAAGAAATGGTAGAACGGAG
CAGCTGGTGATGTGTGGGCCCACCGGCCCCAGGCTCCTGTCTCCCCCCAGGTGTGTGGTGATGCCAGGCATGCCCTTCCCCAGCATCAGGTCTCCAGAG
CTGCAGAAGACGACGGCCGACTTGGATCACACTCTTGTGAGTGTC**FOR**CCC**A**AGT**SINGLE**GTTGCAGAGGTGAGAGGAGAGTCGACAGTGAGTGGGAGTG
GCGTCACCCCTAGGGCTCTACTGGGCCGGCGTCTCCTGTCTCCTGGAGAGG**UNIQUE**CTTCGATGCCCCTCCACACCCTCTTGCTCTTCCCTGTGATGTCATCT
GGAGCCCTGCTGCTTGCGGTGGCCTATAAAGCCTCCTGGTCTGGCTCCAA**HUMAN**G**BEING**GCCTGGCAGAGTCTTTCCCAGGGAAAGCTATAAGCAGCAAA
CAGTCCGCATGGGTCATCCCCTTCACTCCCAGCTCAGAGCCCAGGCCAGGGGCCCCCAAGAAAGGCTCTGGTGGAGAACCTGTGCATGAAGGCTGTCA
ACCAGTCCATAGGCAAGCCTGGCTGCCTCCAGCTGGGTGGACAGACAGGGGCTGGAGAAGGGGAGAAGAGGAAAGGGGAGTTGCCTGCCCTGTCTCC
TACCTGAGGCTGAGGAAGGAGAAGGGGATGCACTGTTGGGGAGGCAGCTGTAACTCAAAGCCTTAGCCTCTGTTCCCATGAAGGCAGGGCCATCAGGCA
CCAAAGGGATTCTGCCAGCATAGTGCTCCTGGACCAGTGATACACCCGGCACCCTGTCCTGGACAGGCTGTTGGCCTGAATCTGAGCCCTCGTGGAGGTCA
AAGCAACCTTTGGTTCTGCCATTGCTGCTGTGTGGAAGTTCACTCCTGCCTTTTCCTTTCCCTAGAGCCTCCACCACCCCGAGATCACATTTCTCACTGCCTTTTGTCT
GCCCAGTTTCACCAGAAGTAGGCCTCTTCCTGACAGGCAGCTGCACCACTGCCTGGCGCTGCGCCCTTCCTTTGCTCTGCCCGCTGGAGACGGTGTTTGTCAT
GGGCCTGGTCTGCAGGGATCCTGCTACAAAGGTGAAACCCAGGAGAGTGTGGAATCCAGAGTGTTGCCAGGACCCAGGCACAGGCATTAGTGCCCGTTG
GAGAAAACAGGGGAATCCCGAAGAAATGGTGGGTCCTGGCCATCCGTGAGATCTTCCCAGGGCAGCTCCCCTCTGTGGAATCCAATCTGTCTTCCATCCTG
CGTGGCCGAGGGCCAGGCTTCTCACTGGGCCTCTGCAGGAGGCTGCCATTTGTCCTGCCCACCTTCTTAGAAGCGAGACGGAGCAGACCCATCTGCTACT
GCCCTTTCTATAATAACTAAAGTTAGCTGCCCTGGACTATTCACCCCCTAGTCTCCATTTAAAAAGATCCCCATGGCCACAGGGCCCCTGCCTGGGGGCTTGTC
ACCTCCCCCACCTTCTTCCTGAGTCACTCCTGCAGCCTTGCTCCCTAACCTGCCCCACAGCCTTGCCTGGATTTCTATCTCCCTGGCTTGGTGCCAGTTCCTCCA
AGTCGATGGCACCTCCCTCCCTCTCAACCACGTGAGCAAACTCCAAGACATCTTCTACCCCAACACCAGCAATTGTGCCAAGGGCCATTAGGCTCTCAGCAT
GACTATTTTTAGAGACCCCGTGTCTGTCACTGAAACCTTTTTTGTGGGAGACTATTCCTCCCATCTGCAACAGCTGCCCCTGCTGACTGCCCTTCTCTCCTCCCTCTC
ATCCCAGAGAAACAGGTCAGCTGGGAGCTTCTGCCCCCACTGCCTAGGGACCAACAGGGGCAGGAGGCAGTCACTGACCCCGAGACGTTTGCATCCTG
CACAGCTAGAGGTCCTTTATTAAAAGCACACTGTTGGTTTCTGCTCAGTTCTTTATTGATTGGTGTGCCGTTTTCTCTGGAAGCCTCTTAAGAGAAGAACACAGTGGC
GCAGGCTGGGTGGAGCCGTCCCCCCATGGAGCACAGGCAGACAGAAGTCCCCGCCCCAGCTGTGTGGCCTCAAGCCAGCCTTCCGCTCCTTGAAGCT
GGTCTCCACACAGTGCTGGTTCCGTCACCCCCTCCCAGGGAAGCAGGTCTGAGCAGCTTGTCCTGGCTGTGTCCATGTCAGAGCAACGGCCAAGTCTGGGT
CTGGGGGGGAAGGTGTCATGGAGCCCCCTACGATTCCCAGTCGTCCTCGTCCTCTGCCTGTGGCTGCTGCGGTGGCGGCAGAGGAGGGATGGAGTCTGA
CACGCGGGCAAAGGCTCCTCCGGGCCCGTCACCAGCCCCAGGTCCTTTCCCAGAGATGCCTGGAGGGAAAAGGCTGAGTGAGGGTGGTTGGTGGGAA
ACCCTGGTTCCCCCAGCCCCCGGAGACTTAAATACAGGAAGAAAAAGGCAGGACAGAATTACAATGTGCCGGCCCAGGGTGGGCAGCGGCCCTGCCT
CCTACCCTTGCGCCTCATGACCAGCTTGTTGAAGAGATCCGACATCAAGTGCCCACCTTGGCTCGTGGCTCTCACTGCAACGGGAAAGCCACAGACTGGGG
TGAAGAGTTCAGTCACATGCGACCGGTGGCTCCCTGTCCCCACCCCCATGACACTCCCCAGCCCTCCAAGGCCACTGTGTTTCCTAGTTAGCTCAGAGCCTC
AGTCGATGCCTGACCCAGCACCGGGCACTGATGAGAAAGTGGCTGTTTGAGGAGCCACCTCCCAGCCACCTCGGGGACAGGGCCAGGGTGTGCAGCA
CCACTGTACGATGGGGAAACTGGCCCAGAGAGGTGAGGCAGCTTGCCTGGGGTCACAGAGCAAGGCAAAAGCAGCGCTGGGTACAAGCTCAAAACCA
TAGTGCCCAGGGCACTGCCGCTGCAGGCGCAGGCATCGCATCACACCAGTGTCTGCGTTCACAGCAGGCATCATCAGTAGCCTCCAGAGGCCTCAGGTC
CAGTCTCTAAAAATATCTCAGGAGGCTGCAGTGGCTGACCATTGCCTTGGACCGCTCTTGGCAGTCGAAGAAGATTCTCCTGTCACAGTTTGAGCTGGGTGAGC
TTAGAGAGGAAAGCTCCACTATGGCTCCCAAACCAGGAAGGAGCCATAGCCCAGGCAGGAGGGCTGAGGACCTCTGGTGGCGGCCCAGGGCTTCCAG
CATGTGCCCTAGGGGAAGCAGGGGCCAGCTGGCAGGAGCAGGGGGTGGGCAGAAAGCACCCGGTGGACTCAGGGCTGGAGGGGAGGAGGCGATC
TTGCCCAAAGGCCCTCCGACCGCAGGCTCCAGGGCCCGCTCACCTTGCTCCTGCTCCTTCTGCTGCTGCTTCTCCAGCTTTCGCTCCTTCATGCTGCGCAGCT
TGGCCTTGCCGATGCCCCCAGCTTGGCGGATGGACTCTAGCAGAGTGGCCCAGCCACCGGAGGGGTCGACCACTTCTCTGGGAGCTCCCTGGACTGGA
GCCGGGAGGTGGGGAACAGGGCAAGGAGGAAAGGCTGCTCAGGCAGGGCTGGGGAAGCTTACTGTGTCCAAGAGCCTGCTGGGAGGGAAGTCACCT
CCCCTCAAACGAGGAGCCCCGCGCTGGGGAGGCCGGACCTTTGGAGACTGTGTGGGGCCCGGGCACTGACTTCGGCAACCACCTGAGCGCGGGCAT
CCTGTGTGCAGATACTCCCTGCTTCCTCTCTAGCCCTCACCCTGCAGAGCTGGACCCCTGAGCTAGCCATGCTCTGACAGTCTCAGTTGCACACATGAGCCA
GCAGAGGGGTTTTGTGCCACTTCTGGATGCTAGGGTTACACTGGGAGATACAGCAGTGAAGCTGAAATGAAAAATGTGTTGCTGTAGTTTGTTATTAGACCCCTTCT
TTCCATTGGTTTAATTAGGAATGAGGAACCCAGAGCCTCACTTGTTCAGGCTCCCTCTGCCCTAGAAGTGAGAAGTCCAGAGCTCTACAGTTTGAAAGCCACTATT
TTATGAACCAAGTAGAACAAGATATTTGAAATGGAAACTATTCAAAAAATTGAGAATTTCTGACCACTTAACAAACCCACAGAAAATCCACCCGAGTGCACTGAG
CACGCCAGAAATCAGGTGGCCTCAAAGAGCTGCTCCCACCTGAAGGAGATGCGCTGCTGCTGCTGTCGTCCTGCCTGGCGCCTTGGCCTACAGGGGCCG
CGGTTGAGGGTGGGAGTGGGGGTGCACTGGCCAGCACCTCAGGAGCTGGGGGTGGTGGTGGGGGCGGTGGGGGTGGTGTTAGTACCCCATCTTGTAGGT
CTGAAACACAAAGTGTGGGGTGTCTAGGGAAGAAGGTGTGTGAGCAGGGAGGTCCCCGGCCCAGCTCCCATCCCAGAACCCAGCTCACCTACCTTGAGA
GGCTCGGCTACCTCAGTGTGGAAGGTGGGCAGTTCTGGAATGGTGCCAGGGGCAGAGGGGGCAATGCCGGGGCCCAGGTCGGCAATGTACATGAGGTC
GTTGGCAATGCCGGGCAGGTCAGGCAGGTAGGATGGAACATCAATCTCAGGCACCTGGCCCAGGTCTGGCACATAGAAGTAGTTCTCTGGGACCTGCAAG
ATTAGGCAGGGACATGTGAGAGGTGACAGGGACCTGCAGGGGCAGCCAACAAGACCTTGTGTGCACCTCCCATGGGTGGAATAAGGGGCCCAACAGC
CTTGACTGGAGAGGAGCTCTGGCAAGGCCCTGGGCCACTGCACCTATCTCCACCTCTGTCCCGCCCCTCCCACCTGCTGTTCCAGCTGCTCTCTCTTGCTGAT
GGACAAGGGGGCATCAAACAGCTTCTCCTCTGTCTCTGCCCCCAGCATCACATGGGTCTTTGTTACAGCACCAGCCAGGGGGTCCAGGAAGACATACTTCTT
GTACCTACAGAGGCGACATGGGGGTCAGGCAAGCTGACACCCGCTGTCCTGAGCCCATGTTCCTCTCCCACATCATCAGGGGCACAGTGTGCACTGTGGG
GTCCCAGGCCTCCCGAGCCGAGCCACCCCAGTCACCCCCTGGCTCCTGGGCTATGTGCTGTACCTGTGTCTGATGCCCTGGGTCCCCACTAAGCCAGGC
CGGGCCTACCGCCCACACCCCTCGGCCCTGCCTTCTGGCCATACAGGTTCTCGGTGGTGTTGAAGAGCAGCAAGGAGCTGACAGAGCTGATGTTGCTGGG
AAGACCCCCAAGTCCCTCTTCTGCATCGTCCTCGGGCTCTGGCTTGGTGCTCACGCACACAGGAAAGTCCTTCAGCTTCTCCTGGGAGGGCCAGGATGGCC
AAGGGATGGTGAATATTTGGTGCTGGGCCTAATCAGCTGCC

ACGGTGAAAACCCGTCTCTACTAAAAAAATACAAAAAAATTAGCCGGGCGTGGCGGCGGGCGCCTGTAGTCCCAGCTACTCAGGAGGCTGAGGCAGGAGAATGGCGTGAACCCGGGAGGCAGAGCTGGCAGTGAGCCAAGATCGCGCCACTGCACTCCAGCCTGGGCGAC
GACTCCGTCTAAAAACACACACACACACACACACACACAAAATCTCAGAGGTTCTCCAGGGTGCAAATAAAAGCACCAGTCAGTTTTATAATAGACTTCGTAAGGCATTTTGGTTGTACACTCCGTTTAACCCTGAGGCTGCTGAAAATCAGCACATGGTGAATACGACATTTGTAAGG
AGGAGATATCGGGCATAAATTACAGAAGTTGGAAGCTCCGCAGGCGTGAATGCTACTCAGCTTATTAAAGTGGCTACCAAGGTGTACATTAACGGAGATCAGGAGGCTGATCAGAGGCTTAAGAAAGGCTAATTTACTAGCAGCAGCGCTTACAGGAAGAAGAAGCTAGCTTTGCAAGA
GACGCGGGCGTGAATGCAGTCGTGGAAAAAGCTAATAAACTGACTTTAGAGACTAAATACCCTAAACCAAGCTACACTGCTCCTCATCAAGTCAGTGCCAAGAGGCCCCCTTCATTGCTGTGTGGACTTGGTAGATAAAGTGTTCTCAAGCCAGAGAGATGTACCAGATCGGCCCCTCC
AGACATTGAATACTCTACTGATGGAAGCAATTTCATACTAAAGGGAGTCCACCAAGCTGGGTACGCAGTGGTGACTTCGGACTCAGTAGTAAAGGTGCAGTCTTTGCCTACAGAAACTTCTGCTTAGAAATCAGAGCTGATAGCTCTGACAAGAACTCCCTGGCTAGGAAAAGACCAAAAC
TACAAAGATTCCAAATATGCTTCTGCCACTTTGCATGTTCATGAGGCTATTTACAAAAAAAAAAAAAAAAAAAAAGCCTTTTAACTACTGAAAGTAAAAAAAAAAAAAAAAGCACAAGGAAGAAATTTTGCAGCTCTTAGATGCTGTATGAGCCCCAAAAGAGGTGGCCATGATGCCCTGCGAGAC
AAAGCAAGAACACCAAAGGCTAAAAAAAATAGAAAGGCAAAAAGGCAAAGCAGGCTGCAGTGACAACTCCACCTTCTAAAGAGGAGGCCTCAGCTATGCCTCTCCTCCCGGAGATTCCCTCCCAGAGATCCCAAGCACCCCTGCAAATAAAAGGGCTTGGTTTGCCCAGAAAAA
ACATTGAAGGAGGACGGTAAAATTCTCCAATGGGAGGCCATACCTGAAATGGTGACCCCGATTTGTAGAACAGTTCCACCAAGGAACTCACATGGGAAAAAAAACAAACAGCACTAAAGACATTATTAAGGCATCATTTCTATGTGCCATGGCTCGCTGCTATTACTCAAGCCATTTGTAAA
GTTTAACTTGTGTTCAGAACAATCCATAACAAGGGCCTGCTTGGCCCCCAGGAGTTCAGGAAACATGAGCCAGCCATGCCCTGTGAAAAGCTGCTTATGGACTTCGCCAAATTGCCCTGAAAGGGGGGCTATCAGTGCATGTGGGTGTTCATTTGCACCTTTTCAGGATAAGTCGAGGCC
GGACAGAGAAGGCACTAGAAGTGACTCAGGTGTTGTTAGGAGACGTTATTCCCAGATTTGGACTGCCTCTGAGATCAGACAATGGACCAACATTTGTGGCTGCAAGAGTTCAGGACTTAACTAGACTATTAAAAATAAAGTGGAAATTGCGTACATCCTACAGGCCACAGAGCTCAAGTAC
GAACCAGACACTCAAACAGCTGCTGAAAAAATTTTGTCAAGAAACTCATCTAAAATAAGATCGGGTCTCGCCCATGGTCCTCCTCCAAGTCAGGTACACCCCCACCAAACAAACTGGGTATTCGTCCCGTGAAATATTGTTCAGAAGGCTGCCCCCAATCATTAATCAAATTAGAGGGGAT
TTAGGAGAGCTAACCCTTAGAAGACAGATGCACTCCTAAAGGAGTGACAATGCAAGAGATGCATGGCTAAGTATTAAAAAAAAAAAAAAATGCCTATAAGTCTAACAGACCCAGTACACCCTTTCAAACCTAGGGTCTGTGTTTAGGTTAAAAAATAAAATCAACCACTCTAGGACCCATA
GCCCCATATTGTGATCAGGAACTGATCAACATGTTCAAACAATAAGAAAATAGTCATATTAAATGTAAAGATTGGTGGGACACAGTGGCTCACGCCTGTAATCCCAGCACTTTAGGAGGCCGAGGCAGACAGATTATCTGAGGTCAGGAGTTGGAGACCAGCCCGACCAACATGGAGAA
TCTACTAAAAATGCAAAATTACCCAGGTGTGGTGGTGCATGCCTGTAGTCCCAGCTGCTTGGGAGGCTGAGACAGGAGAATCCCTTGAACCCGGAAGGCAGAGGTTGCAGTGAGCTGAATTCGTACCACTGCACTCCAGCCTGGGCGACAGAACAAGACTCTGTCTTAAATAAATAA
AAAATAAATAAATAAATAAATAGTATCTTATACATTTAAAGCAACCCGAGGAAGCAAATCAGGGGAGGTTCAAAAAAATGAATAAGGGAGTAAGAATAGGAATAAAGAAAGAAATTGAAATCCCGGTGATTAACATTTTCGTAGCCTGTTCAAATGTAAGTATGTAGTCTATTCTGGGATCACTG
GTTTGCTTTCTACCTCTTAAGCTATTTAAAACACTGAGAAGGCAATATTCCAACTTTCATTTAAATGAAATATTCAGAATTTTGATTAAAATAAATAAATAATTACACAAGTTTACAAGTAACAGTTAAATATTTAGGGACATGTACTAAGTTTTACAAACAGTACAAATATTTAAGAGTGTTGATTGGGAGTAAC
AACTGCCAATAAAGTGGAAGATGAAAGAATAGGACTTTACACAGAGCATATTTAGTTATGGGTCTCTGTCTCCTCCCCACACAGAAAAACTCCCAGACATTCATGACTTCATCCACCCTGCCTGGCAGATAGGCCATATTTCCTGACCCCCTGGCTCTTCACCTCAAAAGGTTATCTTTCCAC
GCAAAGACCCTCAGGACACACAACTCATACTGCCAGCTAAGCATCTACCCCGAGGGACAAGGCAAGCACACACTAGGGCAGCTGGGCCATCCTGGCCCTAATCCCTCCAGCGGTGTCCACACTGAGCATTGCAGCACTTGTAGAAGGTGGTCATCGGCTCATCTGCAGAGCGG
CTGCATGAAGTAAGCACGAGGATGTTCGCATTTGGGACACGACTCTGGCAATGAGAAAAAAGAAGTGACCACATTTAAAAAAAGATGCTACTCATGGCTGTCAAAAGTAGGTATTACTTAATATGTGGGAGGCCAAGGCGGGCGGATCACCTGAGGTCAGGAGTTCGAGACCATCCTC
GGTGAAACCCCATCTCTACTAAAAATACAAAATTAGCCAGGCGTGGTGGCGGGCGCCTTTAATCCCAGCTACTTGGGAGGCTGAGGCAGGAGAATCGCTTGAACCCAGGAGGCAGAGATTGCAGTGAGCCAAGATCGTGCCACTGCACTCAAAAAAAAAAAAAAAAAAAAAAAAAG
CTATAATATCTTTTTTTTTTTTTTGAGATGGAATCTCGCAGTCACCCAGGCTGGAGTGCAGTGGGGCGATCTCAGCTCACTGCAACCTCTGCCTCCCGGGTTCATGTGATTCTCCTTGCCTCAGCCTCCCAAGTAGCTGGGATTACAGTTGCCCGCCACCACACCCGGCTAATTTTTTGTATTTTTA
AGGTTTTCACTATGTTGGCCAGGCTGGTCTCGAACTCCTGACCTCATGATCCGCCCACCTCAACCTCCCAAAGTGCTGGGATTACAGGCATGAGCCACCGTGCCCAGACTACTTAATATCTTTAAAGATGGTAAAGAACCTCCCTGCTCATTTGATCCCCCTTGACAAGGGTAGATCTACA
GTCATTCCCACCACGCTGTAACACTAAGGGAATCTGGGAGACCTGTGACTCACACACTCTCCCCCTCCCCTCCCTGAGTCCCATGGTTCTTCTCAAATCCAGCTGAAGGCTGGTTGGAAACGTAACATGGAAGGTGACTGTTTTGGTCCATTTTGTGCTGCTGTAACAAAATGCCTAAGACTA
TTAGATGGTGGAAGAGTTTTGAAGTAAATGCTGGGAAAAGCCTGCATTGTCAAGAATTGAGCATTAAGGGCCGGGCGCGGTGGCTCACGCCTGTAATCCCAGGACTTTGGGAGGCCAAGGCAGGCGGATCACGAGGTCAGGAGATCGAGACCATCCTGGCTAACATGGTGAAACC
CTAAAAACACAAAAAAATTAGCCAGGCGTGGTGTCGGGTGCCTGCAGTCCCAGCTACTCGGGAGGCTGAGGCAAGAGAATTGCGTGAACCCGGGAGGCGGAGCTTGCAGTGAGCCGAGATCGCGCCACTGCACTCCAGCTTGGGCAACAGAGTGAGACTCCATCTCAAAAAA
TTGAGCATTAAGGGCGATTCTGGTGTGGGCTGAAAAGAAGAGAAGAGGCCGGGTATGGTGGCTCTTGCCTGTAATCCCAGAACTTTGGGAGGCTGAGGCAGGTGGATCACGAGGTCGGGAGGTCAAGACCAGCCTAGCCAAGATGGTGAAACCCCGTCTCTACCAAAACTACAAA
GGGCGTGGTGGCAGGTGCCTGTAATCCCAGCTACTAGGGAGGCTGAGGCAGGAGAATCGCTTGAACCCGGGCAGCATAGGTTGCAGTGAGCAGAGATCGTGCCACTACACTCCAGCCTGGGCAAAAGAGACTCCATCTCAAAAAAAAAGCCAGGCACAGTGGCTCAGGCC
AGCACTTTGGGAGGCTGAGGCCAAGACAGATCACCTGAGGTCGGGAGTTCGTGACCAGCCTTATCAACATGGAGAAACCCCATCTCTACTAAAAATACAAAATTAGCTGGGCGTGGTGGCGCATGCCTGTAATCCCAGCTACTCATGAGGCCGAGGAAGGAGAATCACTTGAACCTGG
AGGTTGCAGTGAGCCGAGATCACACCATTGCACTCTTCAAAGCGATTTCACTTTGAGAAACTTATCCTACACCTGACAGAAATGGCTTCAAACACTCTATTTCTGACATTAGGTAAAAAGATTTTTTTTTTTTAGACAGTCTCACTCAGGCTGGAGTGGCATGGTCTCGGCTCACTGCAACCTCCGC
CAAGCAATTCTCCTGCCTCAGCCTCCTGAGTAGCTAGGACTACAGGCACGCGCCACTGAGCCTGGCTAATTTGTTTTTTTTTTTTTGAGACGGAGTTTGCTCCTGTTGCCTGGGCTAGAGTGCAATAGCACGATCTCAGCTCACTACAACCTCCGCCTCACAGGTTCAAGCGATTCTCCTG
CCCAAGTAGCTGGGATTACAGGCAGGCACCACCACGCTCAGCTACTTTTTGTATTTTTAGTGGAGACGGGGTTTCACCATGTTGGCCAGGATGGTCTCGATGTCTTGACCTCGTGATCCGCCCACATTGGCCTCCCAAAATGCTGGGATTATAGGTGTGAGCCACCGTGCCAGCTGCTAAC
TAGCAGATTTGTATGTTCTTTTTTTTGACAGAGTTTCGTTCTTGTTGCCCAGGCTGGAGTGCAATGATGCAATCTTGGCTCACTGCAACCTCCACTTCCCGGGTTCAAGCGATTCTCCTGTCTCAGCCTCCCGAGTAGCTGGGATTACAGGTGCTGCCACCACGCCTGGCTAATTTTTTGTATTTTTAGC
GGGTTTCACCGTGTTAGCCAGGCTGGTCTCGAACTCCTGACCTCAGGTGATCCGCCCGCCTCGGCCTCCCAAAGTGCTGAGATTGCAGGCGTGAGCCACCACACCCAGCCTATATGTTCTTTTATAGGGGGAAAACCACAGTACTAACCATCAGTTATCTCTCTCACCTCTCAATATAACT
TCTTCCCCTTTCCTCTCCTGGCCACATCTATGTCTACCAGACATGCAATAAACATGTATTTTACCCAATCTGCAGTTTAATGCTCACTGGTATTAACTTCTAAAAATCTATATTAGAAGCGTGTATTAGTCCGTTCTCACACTGCTATGAATACCTGAGACTGGGTAATTTATATAGACAAAAGGGGTTTTC
GACAGAGTCTCACTCTTGTCACCGAGGCTAGAGTGCAGTAGTGTGATCTCGGCTCACTGCAACCTCTGCCTCCTGGGTTCAAGCAATTCTCGTACCTCAGCCTCCTGAGCAGTTGGGTCTACAGGCGAGCGCCACCACACCCGGCTAATTATTTTATCTTTAATAGAGACAGGGTTTCACCA
CAGGCTGGTCTGGAACTCCTGACCTCAAGTGATCTGCCTACCTTGGCCTCCCAAACTGCTGGGATTTACAGGCGTGAGCCACGACACCCAGCCAAGAAAAGAGGTCTAATTTACTCACAGTTCTGCGTGGCTGGGGATGCCTCAGGAAACCTACAATCATGGCGAAAGGAACCTCTTC
AGCAGGAGACAGAATGAGTGCGAGCAGGGGAAACGCAAACGCTTATAAAACGATCTGATCTTGTGAGAACTCACTCACTATCAGGAGAACAGCATGGGGGGAAACCGCTCCCATGATTCAATTACCTCCCACTGGGCCCCTCCCAGGACACCTGGGGATTACGGAAACTATAATTC
GATTAGGGTGGAGACACAGCCAAACCATATCAAACAGAAAGAACATACTTTTTCCTTAGACTCTTATTTTTATTAATCTACTCATCATTGCCATTATCAAAGCTATCACCTGTTGAGTGCCTCTTATATACCAAGCACGGTATAATTCACTTTATGCTCTCGCGGTCTACGAATATGGATTAGCCCCAGA
AGAGACCCAAGGAAGGTCCTACAACAATACAAGGTTGCCCTGAGAATCTGACCCGGGCCTGTATGACTCCAAGGCTGGCACAAATATTGCTTGTCTGACAGATTATTAAACAGTAACAAATACAAATAAATTATTCAAGAACCATTTTCTATTTTTAGAAACCTACATCATTTATATGACATCATAC
AAGCAGTGGCATCAACAGCCTCATAATTATCCACAGTTAGCAACAGAAACAGTTTCCCTCTTACCTGCAGTAGAGTCAACATTCTCCCAGGCAGCTGCTCCACCAAGCACATCATCCACTTCTTTCAGTTTGGGTACTTCCGATTTGTTACCTAACAAAAACAACACTTCAGACTCCAAGTA
GCGCAAACCTGGGCTGCCAAACCTTAAATTTCAGGATTACATAAGAGTTCACTCGGCTGGGCGTGGTGGCTGACACCTGTAATCCCAGCACTTTGGGAGGCCAAGACGGGCGGATCACGAGGTCAAGAGATCGAGACCATCCTGGCTAACACGGTGAAACCCTGTCTCTACTAAAA
AATTAGCCCGGCGTGTTGGCGGGCGCCTGTAGTCCCAGCTACTCAGGAGGCTGAGGCAGGAGAATGGTGTGAACCTGGGAGGCGGAGCTTGCAGTGAGCCGAGATCACACCACTGCACTCCAGCCTGGGCTACGGAGCAAGACTCCATCTCAAAAAAAAAAAAAAAACAAAGA
TCCCGTAGTTTCAAACTCCAAACTCCCACAACAACAGTACAAACTGGAGTAGCATGCAAACATTAAGACTGACGTAGCACCAACATATTCTAACTGCTAAAAAGTAAAGGTATTCTGGCCAGGCGTGGTGGCTCACGCCTATAATCCCAGCACTTTGGGAGGCCAAGGCAAGCGGATCA
TGGGAGTTCGAGACCAGCCTGGCCAGCGCGGTGAAACCCCGTCTTTACTAAAAATACAAAAATTAGCCGGGGATGGTGGCAGGCGCCTGTAATTCCAGCTACTCGGGAAGCCGAGGCAGAAGAATTGCTTGAACCCGGGAGGCAGAGGTTGCAGTGAGCCGAGATTACGCCAT
AGCCTGGGAAACAGAGCGAGACTCTGTCTCAAAAAAAAAAAAAAAAAAAAAAAAAAAAAAAAAAAAAAAGGTCGGGCGCGGTGGCTCACGCCTGTAATCCCAGCACTTTGGGAGGCCGAGGCGGGCAGATCACGAGGTCAGGAGACCGTCCTGGCTAACACACGGTGAA
TCTACTAAAAATACAAAAAATCAGCCAGGGGTGGTGGCGGACGCCTGTAGTCCCAGCTAATTGGGAGGCTGAGGCAGGAGAGTGGCGTGAACCCGGGAGGCAGAGCTTGCAGTGAGCCGAGATCGCGCCACTGCACTCCAGCCTGGGCGACAGAGCGAGACTCCGTCTCA
AAAAAAAAAAAAAAAAAAAAAAACAATAAAAAAAATAAAGAAAGCACACAATACATTACATAATATATTAGAAGGTAATAAGCCCTATGGAAACAATTACAAAGGGAAGGAGAATAAATTGGAAGGAGGGGCTGAAATTTAAAATACGGTAAAGACAAGACCAATGAGTAAGGTGGCGA
AAGAGTGAACTGAGCAACAGCAAGTGCAAAGGCCCTGCTAGCACCATGAGCACGATGAGAGATCGTCCAGGAGGCGGTGTTGATGCGGCAAAGGGCAACAGGAAGGGCATTAGGACTTGAAATCGGAGACGCACGCAGGGGAGGGAGTCAGTGTCGGAACCTGGTAGGCC
AACTCCGGCTTTCGTCTGCGTGAGCTGGAGAAGAGCCGAAGGTTTCTGCGCACAGCACGGACCTGCGTGCCTCAGCTTTAAGGAAATCACCGTGGCCGCCGCTGTGAACGCAGAGAAGGGCGCGAGCGTGGGAGCAGGAACCCAAGGCGGTGGGAAACGGTGGGGCTTTC
GGAAAGTAGAGCCCACAGATCTGCTGCAGACCAGAAAGGGGCGCGAGAAAGAGCGGACAGAGGCAGACGCCGGGGCTGGCGGCGATGGAGCAGCAGTCGGAGGACGCGGAAGGCCTGCGAGAGTCGCCCGCGCCCAGCGCCGGCCTTCGGGTCCCACCTTGCGC
GTGCACGTAGGGGCACGTGTTGCAGGAGAAGCGGTGGCAGCGTTGTCCCTCCTCCACGATCAGCCCGTTCCCGCAGCCGGGGCAGAACAGCAGCATGGTCTCGAACTCCGCAGGCTCCAACTCCCGGCAGCTCCCACTGCCGCTCAGCGCCGATGCGCCGCCCGCCTCGA
TGGTCCTGGCAGCCTTCCCGCCACACCAACCAACCAATAGACAGGGCGATTCTGCGCTCCCGGCCCTGCTGCAGGCTGTCTCGCACTTGTCATTGGTCACTGCAGCCGCCCCACCCCCCCCGGCGCGCCAGTGGCTGGGCGGCCTCGCTGGGGCGGGCGCAGTTCCTGC
GCTTGGCCTCCCTAGTGCGGGCTGGCAGTGCGGGCAGAGCCCGGCTGAGAGGGGCGGCCCTGGAGGAGACGGAGGCCGCGGGTGGGCCCGAGGCGCAAGAGGAAGATGAGGACGAAGAAGAGGCGCTGCCGCACTCCGAGGCCATGGACGTGTTCCAGGAGGGTC
GTGGTGCAGGACCCGCTGCTCTGCGATCTGCCGATCCAGGTGTGTGCGGGCGGGGGCTGGGGGCGCGGGAGTCGTTCCCCGGGGTCCGGGCATCCGGGCGCCGGCAGCCTCCGAAGGTGCTGGTGGGAGACTGCGGCCCAAGGCTGAGGAAGGCGCCTCTGGGCGTA
GTTTTAACCTTGACCGGGCGCACTCGCCTGGGGCGCGAAGCTTAAAGAGGAACCAAAAACCTCATTAGTGAGTGACATCAAATGACATTTTAATGTAATTGTATTTAGGGGGCTCGGGGGACAGGGTCTTGCTGCGTTGCCCGGGCTGGAGTGCAGCGGCGTCCTCCCACCTCAGCCT
AATTGGGACCACAGGTGCGCGCTACCACACCCGGCTAATTTTTAAATATTTGTACAGAGGGAGTCTCACTATGTTGCCCAGGCTGATCTGAAACTCCTGGGCTCAGGCGATCCTCCTGCCTCGGCCTCCCAAAATGCTGGGATTTAGGTGTGAGCCACCGCCTGGCCTGTAATACCTTTA
AGTTGAGGCAAAAAAAGAGCCTACGGTAAAAGAAAAATTTAGTTTAAATAAGAGCAAGATTAGTAGTAGTACTGATTGTTCCATTTGCCTCAAGCGTCAATCATGTGTCAAAGCGTCAGTATCATGCTTTGGGACGATACTGTCTTTTATTATTTATTATTTATTTGAGACTGAGTCTTGGTCCGTTGCCA
TGTAGTGGTGCCATCTCGACTCACTGCAACCTCCCCCTCCCGGGTTCACTACAGGCACGCGCCACCACGCCCGGCTAATTTTTTGTATTTTTAGTAGAGACGGGGCTTCACCATGTTGGGCCAGGATGGTCTTGATCTCCTGACCTCATGATCCGCCCGCCTCGGCCTCCCAAAGTGCTG
GGCGTGAGCCACCGCGCCCGGCCAATTTTAAATGTTTTATATTTGCTCATCATGGATATTTTAATTGCCTTGTATTTCTTAAATGTTGTATTAAAATATTTATTGTTATTACCGTGTTTTGCTGCCCCGGTAAATTTTGCACACCTGCCTTACTCAGTCCCGAAGCCTGGCCTGGGTCCTGATGATCAGGGA
CGTGAACTGACAGGCCTCTACCCACCCCTGGGGGCAGGTCCTCATGTGAGTTTATTCTCTTGCCTGTTTCTTAGCACCTCTTTCCCCCAGAGACTCTGTCCACTATGGACATTAAAATGTGAGTGGAACATGAGATTGGGGCTCTGATGAATTGGGAAACCATAACGTACAGGCAACCACCC
CCTTCGTAAAATGGAGCACCACTTTTTGGCCACTATCGGCTTGTTCTTTGTTGTTGCTCAAGGTGGTAAAGAGAGCCAGGGTTCCCACTTCTGCCCTTGGTCTTTTGTAGGTTACTCTGGAAGAAGTCAACTCCCAAATAGCCCTAGAATACGGCCAGGCAATGACGGTCCGAGTGTGCAAGAT
GAAGTAATGCGTAAGTGCTACCCTCCTCCCTTCAGGTTATGTGGTCCAGGCTTTCACAGCAGGAAGACCTAACAGTGCTGGTCAGCCTGCTCAGAAACTCACAGGCCATGCCCAGGGGTACTGGGGCAACCACAAACTGCCCTGTGCACAGAGGTGTTGGTTCCTTTCCTGCCATCG
TGGCTTTGGGTTCTCACCATGGATCTTCTCCCATCTGTGTCCGTGGTTGCAGCCGTGGTTGTAGTGCAGAGTGCCACAGTCCTGGACCTGAAGAAGGCCATCCAGAGATACGTGCAGCTCAAGCAGGAGCGTGAAGGGGGCATTCAGCACATCAGCTGGTAAGTGGAACAACATTCCC
GCCCTTCGTGGGGCTAGTGCCCTTCTTGGCACTGTCACCAGGCACCACCTGGAAACAGCTCTCAGCTCTGCATGAGTACAGCACCACTGAAGTGATGAGCTCCCTGTCACAAGAGTGATGAGCTCCCTGTCACAGACAGTGCGGGTCGTTCTGTGCCTGGGACTCCTGCCTCGGCCA
CATTCTGCTCTTCCATCGGCATCACCCCATCCGAGCTGCTGGGTATCTTCACTTGGGGACACTGTCGGGAATTTCCAGTGTGTCTGGAAGTGGCCTCCCTAGTTTGGATGGTACACCTGTAGGGGCTCCCATCCCCTTCTCACCTGGGTGCTGTCAGCCCTCACTCTCCTATTGGATCAACTAT
CTGAGTCTCAACACTGTCGCCTGTTGCATTAGCAAGGTTTGTTTGGCCAAGCCGCCCCAGACAGCCCTCTGAGAACAGAGCCTCCTTGTAGCTGCCTCAGACCCAATCTGCACATTGTACAGAACAGCCCAGGTAGGGAGGACAGCTGCCCCAGGTCCCATAGGACTGCATGCCTCA
GTCATGCAGAGCCACTCAGCTCACCCTGCTCAGGGCACGTGGTTTACCTGCATTCCCCTCTTGCAGGTCCTACGTGTGGAGGACGTACCATCTGACCTCTGCAGGAGAGAAACTCACGGAAGACAGAAAGAAGCTCCGAGAGTAAGTGCCGGCCACGTCCTGAGCCGTAGGGCA
CTGTTGCCCCTTAGTCCCCCATGCTCTGCCAGAGGGCTGCGGCTCCCTCTCCGGAGATAACACCCTCCTCCAAGGAGCTTCCTTGGGGCCCTCCCCACTAGGAGAATGGCAGGCAGAGGGCAGCTGGAGGGGCCTTGCTGCTTCCCACCAGTGGCAGCTTAGGGCTACACAGTC
GGGAGGAAATCAGAGTGGGCATGGAGGCGAAAGGCTGTGACTCTATCTGACACTCAGTTCCCTGCCCCGGGGACTTGTGAGTCGGAGTGCTTCTGGGTCTGTGAGAGGATGGCCACTGCACTTCCTGGCCTTCCCAGAAGCCACACGCCTCCCCTCTGTTGTTTCTTCGTGGTCCTCCA
AGGAGGGGAGTCACAGATATGTCCTTCTGTAGGGCAGGTGCTTTATGCCTTTCCTTCTTTTTCAGCTACGGCATCCGGAATCGAGACGAGGTTTCCTTCATCAAAAAGCTGAGGCAAAAGTGAGCCTCCAGACAGGACAACCCTCTTCATCACTGGTGGCTGAGCTTTTTCCCAGCAGGAAT
CGAATCATCGTGCCTCTTTCACAGAAAGGACGTTGTGGTGGCCTCACCCCAGGCATGCCCAACAGTAACTGTCAGCATAAACCTGGGGGCCCTCAGGACTAGGACAGGGTGAGCCAGTGCTCCCTCCTTTCATGTACTTGGCCTGAGACTGACCTCTCCCTAGGTCCAAATGCCCTAG
CAGACCCACGGCTGGCCCACTGTATAAAATAAACCTGTTTGCTTCTTAGTTTGAAAAGTAGAAAGCCACAGTAACCTGGGTAGCAAAGACTGAGATTGCCCCATCACAGAGGTGAGTTAAGGGGAGAGAATTGGTACAGGCAGAGTCCTATAGTCCAAGATGGCGCCACACCACCAAA
GCCACACCACTCCCCAAACCACACAACTGTGTTACCATGATCTCCACAGCAAGGAGGAAATAAAAGCAGAGCGGCTTTAGGGTTTGCATCCTGGAGCTCACAGTGGCAGCAAGCATAGCACAGCTGTGTGTGGGCCCTCCAGCCTCAGCCTTTAGGTTCCCTGCAGCAGGTTCG
GGGAGGGAGTGTGGGGTGGAAATTGGAGCTGCTAAGGGTGGGGGCGGGGTGCTTCCCAGCATGTGGGGAGGGTATGACAGTACATCCCTAGGTCCCTGTCCCAGCGTCGGCTACCAGCCACTCTAAGAAAGAGCAAGGTGCACGTGCCTGGCTGAAGGATTTCACTGGGAAGT
CCCGTGACTGTATCTGACACTCAGTTCCTGCCCGGGTGAGTCAGCCACAAGGGGCCTGCAGGACCTCCCTCCACAGCTTCCAGCCAGAAACAAAGAAGGATAGGGAAGCCCTGGTCAAGTTAATGGCAGCTAAAACGCTCCCAGTCCATTTATTGGCCACATGAGGTGGTCGTCAA
GTTAGAAGGTTATGACAGGAAGTAGTATAATAAATGCCCGGCAGTACGAGGGGTTCAACAGAAGTGAACAAGGCACAAGAAAGAGGTCTGTGTTCAGGAAACAGGCCAGTCCCCACATGGAGCAGGTGACTCCTGTAAGCCTGTGAGGCTCAGGGAGGTCGTGTCTGGCTCTGGCC
GCACACGGCCGCTGGAGCCCGCAGCCAGCTCAGTGGAGCTGAGCGTCCAGTTCGTACTTCTCACAGAACTTGTCAGTGAAGGGGATGCAGGTGAGGAACTCACACCACTCACAGCGGACAGGATAGACGTAGAAGAGGACCACCAGGCCAGCCAGGAGGCCCAGGAAGAC
AAGATGATGATCTGGCAGCGTTTCCGGTACAGGTCGAACTTGCCAAAGCTGATGTAGGGCAAGAAGGCGAAGGAGAGGAAGAGGCCACTGATGAACCCTGAGATGTGGGCAAAGTTGTCAATCCACGGCAGCAGCCCAAAGGTGAAGAGGAAGAGCACCACAGCCAGCAGC
GGCACGCCAGGGCCGCGCCAGGATCTGCCAGCTCTGGAAGAGCTCCACGAAGAGGCAGGCCAGGATGCCGAACTGGGAGCCAGCAGGACCCACCTGGGGGATGGTTGGGGTAGCTGTAAGGCAGTGGGGCTGGCAGGCCCCCTGGACCCAAGGCCTCATCTTTCTGCCC
CATGGCCCAGAGCATATTGGGACCGAGGATCCAACTGGAGATGCTCCCAGACTCAGTGCTCGATATGGACAGGTTCACAGCAGGCTGCAGGTGCACACGGGAGGATCCCCACCCTGAGGGCCAGCCTTACCTCTGCTCGGTATGGCAGGAAGATGGCACTGGCCAGGTTGCCG
CACTCAGCAGGTAGATGATGGCTATGCGGTGCCAGCCTGCCAGCTTCTCCAGGTCCCGCAGGACAGTCATCTGGAAGCAGATGGACACCAGGCAGTGCAAGATCCTGTAGTCAGTAGCAGGCGGGGGTCGGGAGACATTCAGCAGGGAAGTGACCTTTCTGCCACCCCAGCC
CCTGAGGGCCCTGAGACCCCCGACCCGGCCCACAGTGTCACTTGCCCGGCGTGCAGGAAGAGGGATAGCCACAGGCGGTAGAACTGGTCAGGCACCTCGGGGTGAGAAAAGGCAGGAGCCCACACACATCATCCATGCAGTGCACCTGCGCAGAGCAGCATATCAGCA
GCTGGGGGAGGGGAACGACGGACACTCTGCAGACCTACCTGAGAGCAGAGCGTGGCCTCCTCATGGAAGTAGCCCCTCATGAAGTCACAGTACTCCCGGGAGGTGATCTCACACCTAGAAAGGCAGGCCAGGGTTCGGAGACTGCTGCCTTGCAGAGGGGGATTGCTGGCAT
CGAGTTTCAGCCGCACCTACCCACCTTTGTTGGCCCCAAGGAAGTCCAGACCCCCTCTAACCCCACATCCTTGGTATACTGGCTTATTTCAGGATTTGGTTAGGGAGAAGGCACACCCTAGGGCGATGCTCTGAGCCCAGAGACCAGCTGCACTCGCTGGCACGCACACGTACCTGC
CCAATGCAGCAGGGCCGTCCTGTGATGACACAGTCCATGTGGGGATGGTTGGTGTGGTTCCCAGCGCTGTTTTTGGTGCAGATCTGGGTCACAAATGAGGACGAGCTGAGTCAGGGCCTCCCCCAACCTTTGGCCCCACACCACCGTTTGGGGGTAGAGGCAAAGATGGGTGGGCTC
AGAAAACGGTGTGGCTCCTTGACTTCAAGCAGGCCACCACACTGAGTCTCTATCAAACTGGAATAATGTCACTTGACCTTTCAAGTCTCCATCTAGTTCTCTTGATCTACTTGCTCAGAGGGGCAGGCGCTGGTTGATGGACAGGCTGTACAAAGGGCAGGTGCACAGCCTCTGAAAACG
TCCCATGATGGGAATCTCTGGTATCACCAGTGGGTGGGGTGTGGCATTCTCTCCCACATGACACGGAGGGCTCCCTAACAACCCCCCACTGCAGCTCACCGGCCACTTGGTGATGTCTTCTGGCCACTCATGAGGGTCTTCGGAGGAGGGCTCATCACACACCCTGCATGAGCCA
TCAGACCGGCTTCGAGCCCCTAGCATTGGCACCCTTCCAGCCAAGTCGCCAACAAGAGGAGCCCCTACACCCTCTGCACCACAGCCCCCGGGGAAGGTGTGGACAAAGGCAGGTGAGAGAGCTGCCGGCAGGACTAACCTGGGATCCTGGTGGCAGACAGAGCCAAACT
GGCCCGCAAGCTCTGGGGCGCTGGGATGGATGGGCCACTTCACCCACACTGCCAGCGTGGACTGTGGGAGGGGGACACAGGGTCAGGTCCAGTTGGGTCAGGGCAATGAGGGGCACGCAGGGAGTCAGGAGTCGAAGGCAAAACCACTGAGCTCTTCCCTCCGCTTCTAC
CCCCACTCTTGAGATGCACGTGCATGTTGGAAAAGACTCTTACAAGGAGTTTGACGGCATAATTCCTGTTAGCTGAAAATCAGAAACAACCTCTGTGTCCTAAATCAGGGAACTGGTATGGACTACGGCACTTCCACATAAGATGATGTGATGAAAGTTGGGATGAATTTTTTAAAAATTGCTTAC
GAGGCAAATACCACTTATGTAAGCTTTAAGACACATGCTGGAAGAGTACACCCAAATCAAAAGACCAAATCAGATGATGGTCTACGGGGCGAGGAGCTGGAGATGGGGGAGGGTGGGGATGAGGAGGAATGAATATGACAAGGCCTTGCGCGTAAGGACGGCGGGATGTGCCAC
GGCGAAGGATCACTCTGCCGAGTCTATCTGAGGTCTGAAGAGCGAACACCGGGCAGACGAAGGCGGGAGTCCCGGGCGGGGAAACGCACCGAGCACTCCTCCTCCGAGGTCTGCACGCAGCCCGACCTGTCGTTGCGCACGCAGCAGGCGGAGTGCTTCTCGCGCTCGC
GCGAATGAAGCTGTGCACCTGCGGGTCCTGGCGCATGCAGGGCGAAAACTTGGCGCCCAGGTGGATGAGGGCCTCCTGCGGGCGAGGGAGACGAGCGGCCGCAGTCCGGGGCCTCCTGCCCCCGCCGGGCACCTCCAGGTCCCGCCCCCGCCGGCCCCGCCCCCA
CCTCACCGAGCTGGGCCCGATCCAGAAGTTCTCCTGCTGCACGTACTTGACGTTCTCGTAGACCCCGCGGTTCCGCAGCACCTGGGAGGTTGGGGAGCGGGATCAGACGGGCCCCGACTCTGGCCCTCTCCTCCCAGAGCGGGTCGGGAGGGGGTCCAGTGCCCGGACTCC
GGGCCCAGGGAAGGCACAGGGCTCACCGAGTCCACCGTCTCATGCTGCGAGAAGCCCACGGGCGCGATGCCATAGATGCACACGGCTAGGATGGTGACGAGCGAGTGCACGAAGGTAAGCCAGTAGGTGAAGAAGGGCCTGCGGGGTGGAGCGTCAGCGGGGCC
GACCCGGAGCCCCCACCCTCCCCCGGGAGCCTCCATCCCAACCCAGGCCTTGCCTGCCACGCCTACCTGTGGTCGTCCATGTCCTCGATCTGGCGCTTGACGAAGCTGTCGATGCGCTTGCGGTAGGTGCGGTTGGTGAGCCGTCCCACCATGCCCAGCCCATACGGCCGCTT
GCGAAGAGCTTGCGCACCGGCACCGCGATACGCTGGCCCCGTCGCGGCCCCGCGGTGCTCACCACCTCCTGTCGGAGCCGCACCTTGGGCTGCGGGGCTGCGGCGCCCTCCTTCTGCTTCCGCCAGCCTCGCTCCAAGGGCCTGGAGGGAGCCGGCATTCACCTCCCGC
CCGCAAACTGCTGCAGTCCCTCCCCGGGGGGCACCAGGGCACCCTGTCTCTATCCTGGCGAATATACTGCGCAAGTCGCCCGGCATCTGCCTTCTCCTTGCCAGTAAAAAAGCAACACTGCGGGCCTCCGGGGATGGCAGGGAGCTGGAACGGGGACACACTGACTTCTGCAA
AGTATCACCTCTGTCACCTCCGCCTCCCCGTTGGATTCTGCCCAGCCCCTGGGTTCTAAGATTGCTGTCCCATTCCACAGGCAAGCCATGAGACTGGGGCATGCCCAGCTCTGCGTCTCCCACCCCTGGTGGCCCAGTGAGTAGTGAGTTCCCAGTCCCTGAGGCTACCCCTTCCATC
GAGGTGAATGCCGATACTCGCCCAGCTGTTCCTGACGGGCCCCGGGGTCTCTCTCTGCCCCTCTTCCCTGCCTGCCGCACTCACAGCATCAGGTGGCTGCGCTCAAGCTCGCTGCGGTCCAGGGCCCCGCCGGTGAGGTCCGCCTGCTCCGGTGCCTTCTCCCAGTCCTTTAGCC
CGATGGGGACTCGAAAACTTCATCCGGGTATGTGGACAGCTCTTCATGGAGGATACCTTCTGAGCAAACACATGAGGTCAGGGTGGACACAGGCAGGAAGGAACGTGGGGTGGGGGGACACCCGGGCACTTCTTACCCGGGCAAAGAAGGATGTGTCCAGCTCATCGGGG
TGTGTCCTCCTCCAGAAAGCTAGCTGGAGTGAAGCTTCGACGCTGTGCCCGGCGAAAGGTGCCATCCCTAACGGAGCGGCCCTGGGGACGGGGAAGTGAGCATGAGACCCGGCTGCTGGGTCGGCCCCCAACCCCCGCCTTGGCCCCTGCACCCACTTTCATCAGCGCTGCC
CGGAAGCTCATCTTGGCCACCGACTCTCGCTTGCGCCGCCGCGGGAGCCGGTGGAAACCTGAGCGGGAGCTGGAGAAGGAGCAGAGGGAGGCAGCACCCGGCGTGACGGGAGTGTGTGGGGCACTCAGGCCTTCCGCAGTGTCATCTGCCACACGGAAGGCACGGCCA
AGGGGGTCTATGATCTGGAGGAGGGGAGGAGATGCTGGAGTCAGGACCATGGGGGCTCCTAGCTCACCCAGAGGCACAGAGGCCTTGCAAAGACAGCGTGACTGCTGCTCCTGACTCCTCACGGGCCCCTGGACAGCTCTATGGCCACCACTCCCACCCCACAGCACCTAGA
CCTGGAATGCTCTCCCATCCAGACACTGGCCCCCTGGTGCAATCCCAAGACCATGTCCCCATCCCAAGGTACCCCTTCCCAGAGACTGTTTCAGGTGCCTTCACGAGCAGGCTGTCTGACTGCCCTTCCAGACTGTGAGCTCCCAGAGAGTCCCTTCTGCTCAGGCTGGCTGTGTCCC
CAGCCCCTGAGAGCGGAAGGAGGGGCCTGCAGGAGGCAGCAACAGGCAGGCATACCTTCTGCATGCCCAGCTGGCATGGCCCCACGTAGAGTGGGGGTGGCGTCTCGGTGCTGGTCAGCGACACGTTGTCCTGGCTGGGCAGGTCCAGCTCCCGGAGGACCTGGGGCTTC
GTAGCGCTGGCTGCAGTGACGGATGCTCTTGCGCTGCCATTTCTGGGTGCTGTCACTGTCCTTGCTCACTCCAAACCAGTCGGCGGTCCCCCTGGCATGGCAGGGACTCAGCAAGGGGCGGCCAGATCGCCCCTCCCAGGCAGGGGACTTCAAAGGCCCTCCTCCGGGGCAG
CCCCAGGCCTGTAGTCACTCCAGGGACCAGGGTCTGAGTGGGCAAGGTCTAATAAGAGGATGAGTGGGTTCCAGCCACCCCCTGAACTGTTAGCCAACTCCAAGTTTCCACTAGACCGCAGCTGGCCTTGCCCGGTTGAGTGCTTCATAGAGGCCCTCACCCCACCATCCCTGAGG
GAGGGGAGGCACTGAGTCCAGAGAAGGAGCTGAGAGCTCAACCCGAGAGAGCTTGGCCCCAGGACACCACAGAGCTCAGACAGCAGGTGCTTACGGGCCCAGGCAGGCTGTCCACAGCACTTGGACATGCAAACAAAAACACTGCCAGCACCTGCCGTTAGGAGCCCCA
CCAACCACACTGGCAGCTCCCAAGGAGGTCACATGCCAAGCACATGCCAGAAAACACCAACACACTCATGACCTGTATGAGATACCTGAGCAGGGACCAAGAGAGACTGGCGCTGTGGGAAGGCCCTGCGGGCAGCCCGGGGACCCAAGAGGGGCAGAGTGTGGACA
AGCGTACTTGTCGGCTGCTAAGAGGCAAGAGCATCCGGGCGGTGGAGACAGAGAAGGAGAGAGTGAGTGCTGAAGAGAAGGGAGTGGAGCAGGGTAGTGGGGTGAGAGCGGTCAGTATCCAGAACAGGAGGAGGGCAAGGGGATGGTGGGGACCGAGATGGAGGAGC
ACAAGAGGGCCCTTGTTCTCATTTCCATCCTCTGTGTACCTGCCTCTGCCCCACCAGCCCCCAGCTCCCACCCCATGGAGCAGCTGGGCGGTGTTGGTACCTGCGGATGGTCTGTGTGATGGACGTCTGGCGTTGCAGCACCGGCCGCCGGAGCTCATGGTGGGGTGAAGAGATG
CTCGGCTGGCATACTCACACTCCTCAGGAAAGCCTGTCGCCTCAGGGGCTGGGCAACAGGGTCATGGTGAGGGTACTGGACAGGGGGCACCCACAGTCCCTGCAGGCCAGGGCCTACCTGCAGGAAGCTGGGCTCTTCTGCCGTCAGGGGCACCGCAGAGGGAATGTCCAC
AGGGTGGCTTCTTGCGCTGCAGGCTGCTCGTGCTGTCCCTGCGGGCCTCACTCATGGTTCCTGGCAGAGCAAGGCAGGCCTGCGGGGCATGGTGTGTTATTCAATAATGACAAAGCTGACATCTGAAGATTCATTGCACCCAGACCCGGGAGGAGGGAGAAGCAGTGATGAGCCCA
TGCTCATGTCCCCCACTGTCAGAGCACTACTGTGTGCTCAGCACCGACTTTCTCTGTGCCTGCCTTGTTCATCCCAGCAGACAGTGCTGGGCACAGCCTAGGTCCTCAGACAACATAAGTCCATGGGACAACTATCTGAGCACACACTATGGTCAGGATACTGCTCTAGGTGCTTGAGGC
GGAGAAACGAGACAAAATCCCTGCCCTCCTGGAGCTTACAGTCTAGACAATATTTTAAATGAGTGAATGTACTGGTTAAAAGATGCTAAGTATGATGGGAGCGGGGGAAGCAAACTAGGGTTGCTTGACAGGAGGCAGTGAGGCCTCCCCATGAAGTAACAGCTGAGCAAAGGCTTG
GGAGTGAGCACAGCTGGGGTTGCAGGGTTTCCACACACAGGGGACAGCCGATCTGAGGTCTCAGTGACAGGAGGCTCACCCTAGTCTGTGCCCAGGGCAAATCTCACACCTGTCTTACCAGGCCTCACCCTCTTACCCACCCAGACTTCCACTCCAGGGACAAAGGGCCCAC
TGACAGCCTGCAAGAGGTGTGTGATCGGTCCAGTGCATAGAGTCACCACATTCAGGAAAGAGGGTGACACAGCAGCAGCTAGTGGCCCCAGGCTCCTGGGCACAAATGACTTTCCAGAAACCACCCGGGCGCAGACACCGTAGTGACCACCAGGTGACGCCACGCGCCCAC
AGTGAGGCGGCTCGGTGCCGGGGCTGCGGGGCATTGCAAGCTTGGCTCTACCATGACATGCCAGGCACGGGGCCAGGGCCCCACGGAAGGCCCCAGAGCAGATAGGCCAGGGTTCGATTCTGGCCACCACTGCCTGGCCATATGACTACGACTCCATTTCCCCCTTACACCT
GACCTGCAGCTGTCCCCAACCCAGGGGGACCACACGGCGATTCCAGTGGAGACCCAGATTGAGAAGGGCCCCTGAGCATGACACGAGACAGATAGTGGGGGGCTGGCCAGAGCTGAAGCAATTTTGGACAGAAAACCCCAAATTCCTCAGCCTCTTTAAAGAGCTGCAGCCT
CTCGGAGCCGCTGGGTCGGCAACCCACCCACCAAGGCCCCAGAAATGGCTTTCACTCACAGGTCTCCCTCCCAACTGCTTGGAGGAAAGCCCCACCTATGCTGTACAAATAGGGAAACTGAGGCCAGGAAACACAGGCTTGCCTCGCTCACGGTCATATATGGAGGGCAGCAC
ACAGCCCAGGGTGCTGCTCACAGCTCCTAATGGTCTATGCCAGCCAGGGACAGGCCTGGGACAGAGCTCACGGCTCCTTCCGACCCTGTACTACCAGGGAAAGGCTGGGAATGCTGCTCACAGCCCCTACTGGTCCTACAGGGACCAGAACAGACACAGGTGGAGGAGGGG
GATGCCTTCTGCTGGCTCTTACCCCACCCCTCACGCTCCAGCCCCTTATCTGAGCTGACTGTGCACTTCAGGTCCCAGGTGAGAGCTGGGGGGCAAAGCTTGGGGCACTCCCCAGGCATGGTGCCCAGGGATCATTCCATCCACATGGCCCCCTAAGGCTGAGTTCCTGTCCCTG
GGCCTGGAATCAGGGGCACGAGCCAAGGGCAGCAACTCCACATCACTGGGAAATCCCTCCAGGGCCCCTCACCCATGCAGAGCCCCAGAGTGCAACAGGGTCGGACAGAAAGGCATCCTAACTGAGAACTGGGTGGGTCAAGGTCCCTCTGGGAACTGCACCAGGAGA
ATCTGACTCGTGACCCAAACCCAGCCCAGACCTGCGTGTCCTGGATACGGCCCAGCCACAGCGTGAAGGGGTCACAGCTAGACTTCCTCATCGTCCCCCAGCCCTCCCTCCACAGGGTATATGGACTGCAGTACCATCACCCTGCTGGTAAGGAATAGCCAGGAGGGGCCAC
TCAGACACTAAGGGTGGAATTTCAGACAGCATCAGACACGTTGGGAAGGCAGGTGGTGCTCGGCAAGGGGCTTCCGGAACCCACAGTGCTGGGACAGCATTCTGAGTACACTCCCAGCATGTTCTCACCCAGAAAATGGACGAATGCTCTTAGTGAGGCCATTATGGGGGCCAGA
CAGGGCCCAGCGTAGAGACTGGGGCCACTGCCAGGAATCAGCAGGCTGAATCCAGGGGGCCAGCCTCTCCAGGAGTCTCCACCTGTGCCCAGACACTAGGCGACATCCAGCTGGTTCTTCACTCAGGAGTGTGAGTGACCTTTGCCAACAGAGACCGTTGCCAGCTGATGC
CACATCCATGACTTCCCTGGCTCCTGCTCTCCGCTGCTCCTCACACCTGACTCCCTCCCATCTCAGAGCCCTGCAAATGCGCTCTGCCCACCCTCCCCTGTGTGGGAGGACCTGCAGCCTCAGGCTGCCCTACTACCCAAAGGCGGTCCTTGGCCTGGATCCCTGGGCACAGCACA
AGTTCATTCACATATGTATGGGCCACTGTCCTGTCACTGCTGCCTCCACAGTGCTCAGGACTGGCCTGCTCTGAAGCCACCGCTGGGTAGGCACTGATCACTGAACAACGCAGGCCATCGGGGTCCCTAATCTGGGGTAACGGAGGATTCCTCAAGATGACCTCTCAGCTGGGTCTG
AGCTGCCTGGCATGAGCGGATGCAAGCAGGAACAGTCAGGGGGTGAGAAGAGCCGAGAGCTGTGAGCGTGTACGGGGAACCTGACCTCCCCAGGACTGTTTCGGCCGGGGACCACAGCCGGCCCCAGGAACCTCCCCAGGACTGTCTCAGCCGGGGACCACAGTCGG
AACCTCCCCAGGACTGTCTCAGCCGGGGACCACAGTCGGCCCCAGGAGACAGCAGAGTGCTCAGCTCATGAAGGAGGCACCAGCCGCCATGCCTCTACATCCAGGTCTCCTGGGGTTCCCACCTCCACAAAAACCCCCACTGCTAGGAGTGCAGGCAGGAGGGGACCTG
CAGTTATAGGTCCTGCGGGTGGGCAGTGCTGGGTGTTCTGGTCTGCCCCACCCCTGTGTGCCTAGATCCCCATCTGGGCCTCAAGTGGGTGGGATTCCAAAGGAAGAGCCGGAGTAGGCGTGGGGAGGGGCAGGCCCAGGCTGGACAAAGAGTCTGGCCAGGGAGCGGCAC
CCCAGAGACAGTGGCTCAGTGTCCAGGCCTTCCCCAGGCGCACAGTGGGCTCTTGTTCCCAGAAAGCCCCTCGGGGGGATCCAAACAGTGTCTCCCCCACCCCGCTGACCCCTCAGTGTATGGGGAAACCGTGGCCCACGGAAGGCCTCACTGCCTGGGGTCACACAGCAT
TGCAGCAGCCTCACAGCTGCAGCCAGCCCAGGCCCAGCCCCATCAGGAGACACCCAAAGCCACAGTGCATCCCAGGACCAGCTGGGGGGGCTGCGGGCAGGACTCTCGATGAGGCTGAGGGACGAGGAGGGTCAAGGGAGCCACTGGCGCCATGCATGCTGACGTCCCC
CCTGCAGAGCCTGGTGTGGAAGGGCTGAGTGGGGGATGGTGGAGAGTCCTGTTAACTCAGGTTTCTGCTCTGGGGATGTCTGGGCACCCATCAAGCTGGCCGCGTGCACAGGTGCAGGGAGAGCCAGAAAGCAGGAGCCGATGCAGGGAGGCCACTGGGGACAGCCCAGG
GGGCCCCATGTGTCTCCACCACCTACAACCCTAAGCAAGCCTCAGCTTTCCCATCTGGAAATCAGGGGTCACAGCAGTGCCTGGCACAGTAGCAGCGGCTGACTCCATCACAGGGTGGTGTAGCCTGTGGGTACTTGGCACTCTCTGAGGGGCAGGAGCTGGGGGGTGAAAGGA
GCATATGCAACAAGAGGGCAGCCCTGGGGACACCTGGGGACAGAACCCTCCAAAGGTGTCGAGTTTGGGAAGAGACTAGAGAGAAGCTCTGGCCAGTCCAGGCATAGACAGTGGCCACAGCCAGTGGAGAGCTGCATCCTCAGGTGTGAGCAGCAACCACCTCTGTACTCA
CTGCACACTCACAGGACCATGCTGGCAGGGACAACTGGCGGCGGAGTTGACTGCCAACCCCGGGGCCAGAACCATCAAGCCTGGGCTCTGCTCCGCCCAAGGAACTGCCTGCTGCCGAGGTCAGCTGGAGCAAGGGGCCTCACCCCGGGACACCTTCCCAGACGTGTCC
CATGAGCCTCATCCCAGGGGGATGTGGCTCCTCCAGCATCCCCACCCACACGCTGCTCTCTGACCCTCAGTCTTCTGTTTGACTCCTAATCTGAAGCTCAATCCTAGATCTCCCTTGAGAAGGGGGTCACCAGCTGTCTGGCAGCCCAGCCTCCAGGTCTTCTGGATTAATGAAGGAA
GGCCTCTCTGCCTTGTCTATTAATGGCATCATGCTGAGAATGATATTTGCTAGGCCCTTTGCAAACCCCAAAGTGCTCTTCAACCCTCCCAGTGAAGCCTCTTCTTTTCTGTGGAAGAAATGAGGTTCAGGGTGGAGCAGGCAGGCCTGAGACCTTTGCAGGGTTCTCTCCAGGTCCCCAG
GACTGGCACCCTGCCTCCCCTCATCACCCTAGACAAGGAGACAGAACAAGAGGTTCCCTGCTACAGGCCATCTGTGAGGGAAGCCGCCCTAGGGCCTGTAGACACAGGAATCCCTGAGGACCTGACCTGTGAGGGTAGTGCACAAAGGGGCCAGCACTTGGCAGGAGGGG

Sixty thousand and thirteen

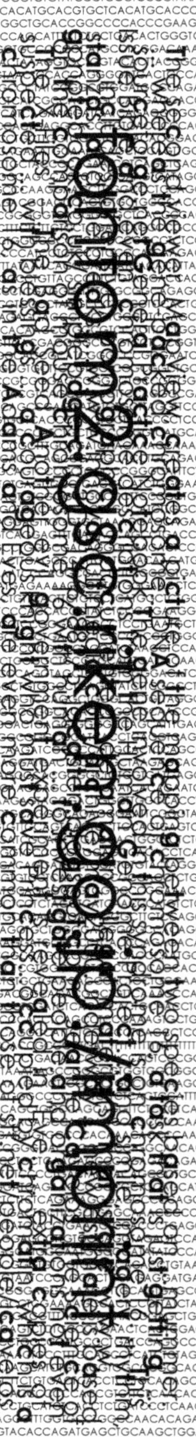

One hundred and ten thousand two hundred and ninety-four

NATURAL ANTI-SENSE

F

7

CLICK HERE TO JOIN NOW!
R

One hundred and ten thousand four hundred hundred and sixty-nine

REALITY IS WILDER

Two hundred and ten thousand two hundred and ninety-six

BOFFINS DEVELOP HUMAN SKIN PRINTER

F

Two hundred and fifty-nine thousand seven hundred and fifty-five

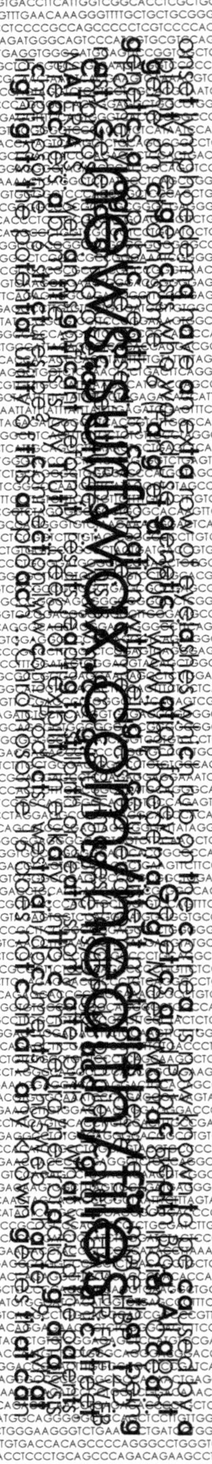

TRACK NEWS SAVE/EXCHANGE INFORMATION

N

Three hundred and nine thousand two hundred and ninety-eight

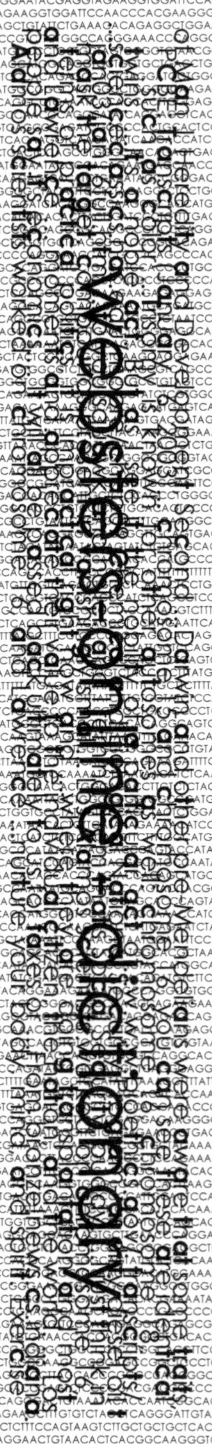

CHROMSOME CHROMISONE CHROMZONE

W

Three hundred and fifty-nine thousand one hundred and ninety-two

A NATURE AND NURTURE WALK

Four hundred and nine thousand one hundred and sixteen

HOMO SAPIENS UPDATED! MUS MUSCULUS UPDATED!

E

Four hundred and fifty-eight thousand two hundred and forty-eight

P

PROTEOME ANNOTATION KNOWLEDGEBASE

Four hundred and seven thousand six hundred and ninety

HOTLISTS STARS GENRES

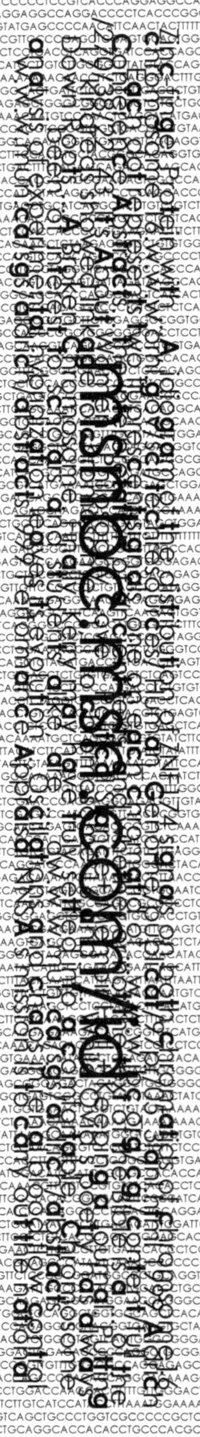

Five hundred and fifty-seven thousand five hundred and forty-two

A CHIMPANZEE NAMED CLINT

M xii

Six hundred
and six
thousand
eight
hundred
and five

SWISS POP & ROCK

Six hundred and fifty-six thousand one hundred and sixty-four

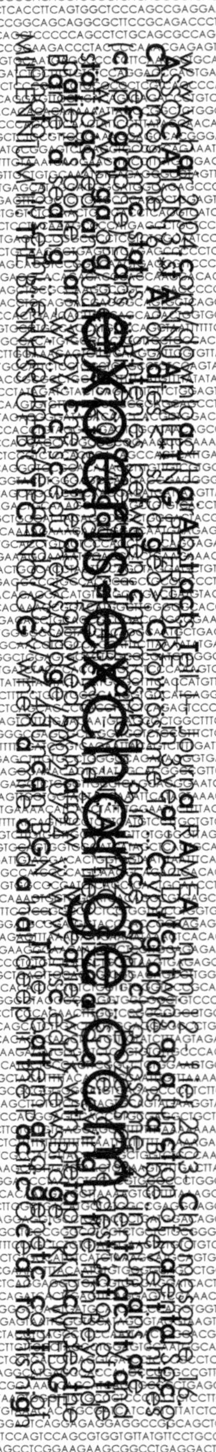

WHY IS IT SO PERFECT?

Seven
hundred
and
six thousand
and ninety

WHAT RELIGION ARE YOU?

A

xxv

Seven hundred and fifty-five thousand four hundred and fifty-six

ALEX THE FAMOUS TALKING PARROT

B

xii

Eight hundred and five thousand five hundred and ninety-eight

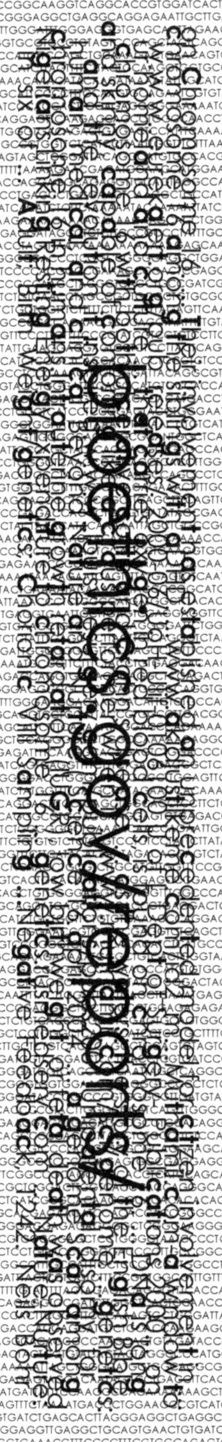

THE PURSUIT OF PERFECTION

B

Eight hundred and fifty-five thousand eight hundred and eight

Erratum. human-nature. com has inadvertently been omitted.

THE RISE OF THE OCEANS

Nine hundred and six thousand eight hundred and ninety-seven

IF YOU LIKE JIGSAW PUZZLES

P

Nine hundred and fifty-seven thousand seven hundred and ninety-two

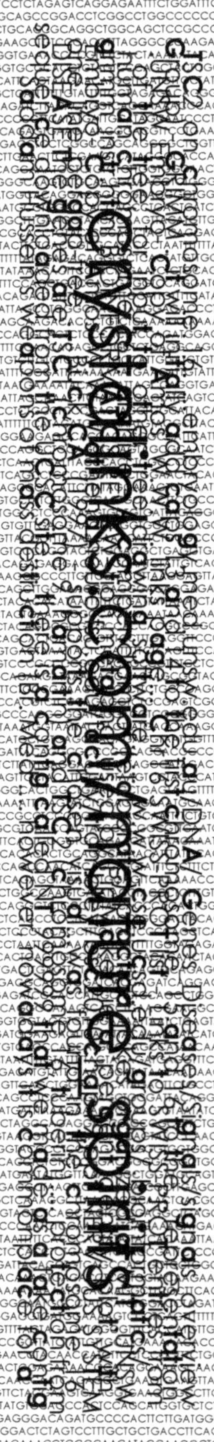

READ THIS FILE WITH MUSIC
C
xxi

One million nine thousand three hundred and ninety-six

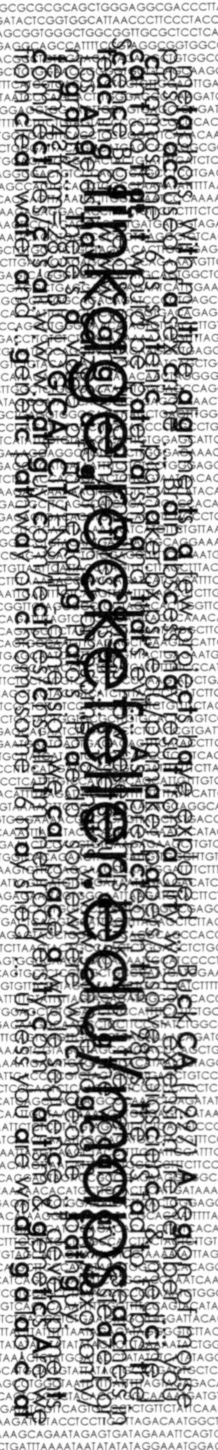

WHEN A LIKELIHOOD RATIO IS GIVEN

L_Y

One million fifty-nine thousand seven hundred and twenty-six

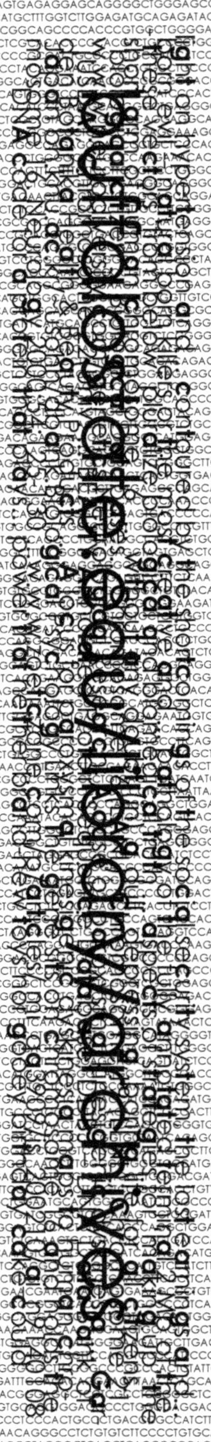

A CLASS ON ENSEMBLE WRITING

B

xvi

One million one hundred and eight thousand five hundred and forty-three

RIGHT FROM THE START I NEED TO MAKE IT VERY CLEAR

One million one hundred and sixty thousand and fifty-seven

THE FUCKING OSCAR RESULTS

One million
two
hundred
and eleven
thousand
three
hundred
and ninety

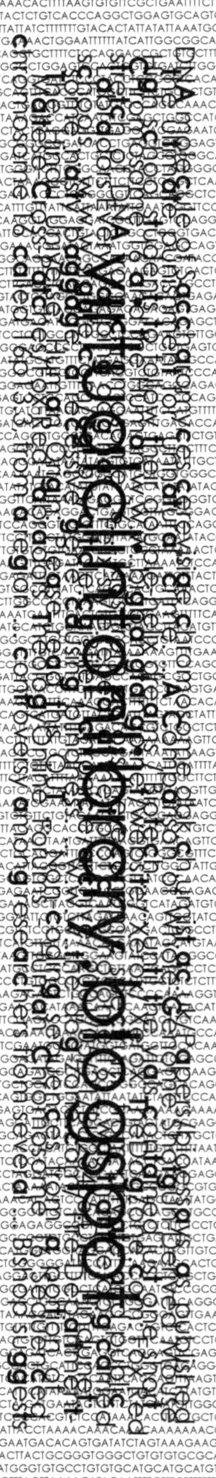

MUTABILITY OF MEANING

V

One million
two
hundred
and sixty
thousand
four
hundred
and
seventeen

A ONE-SENTENCE OVERVIEW

S

xiv

One million three hundred and nine thousand five hundred and thirty-nine

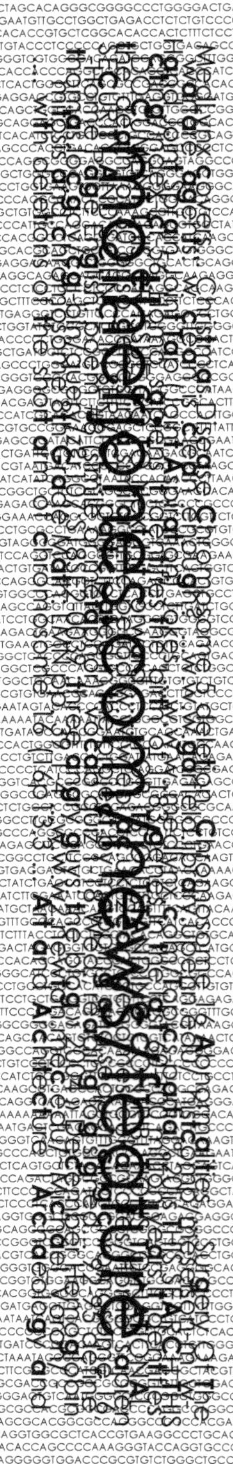

THE PROMISCUOUS OVERUSE

M

One million three hundred and fifty-nine thousand seven hundred and ninety-eight

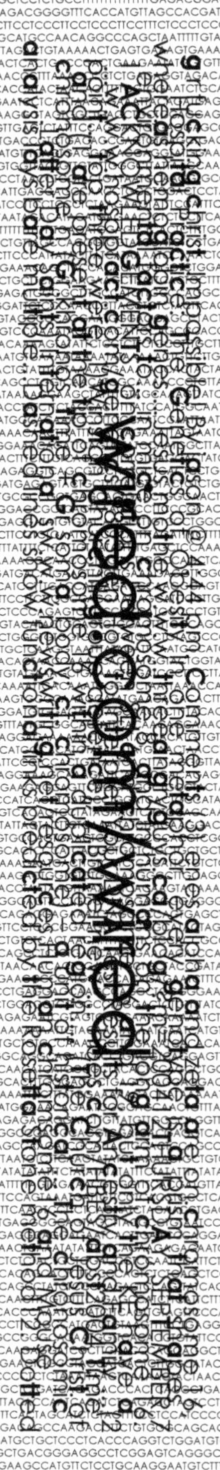

YOU'RE WRONG IT'S THE OTHER WAY ROUND

W

One million four hundred and eleven thousand five hundred and thirteen

COMBINED ANALYSIS OF DATA

T xvi

One million four hundred and sixty-two thousand seven hundred and forty-six

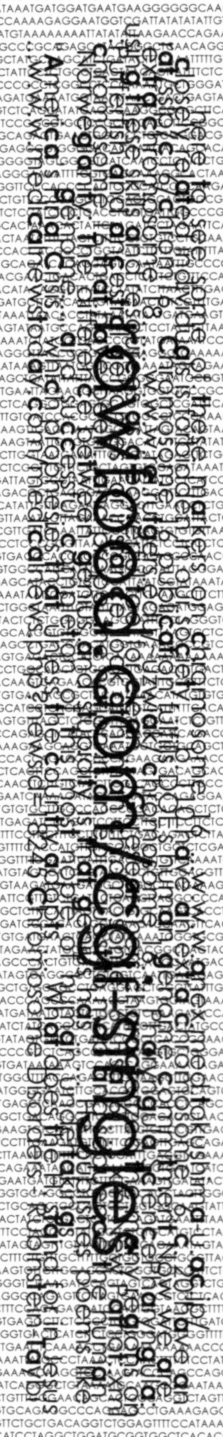

EDIT OR DELETE YOUR PERSONAL AD

R

One million five hundred and thirteen thousand one hundred and ninety-nine

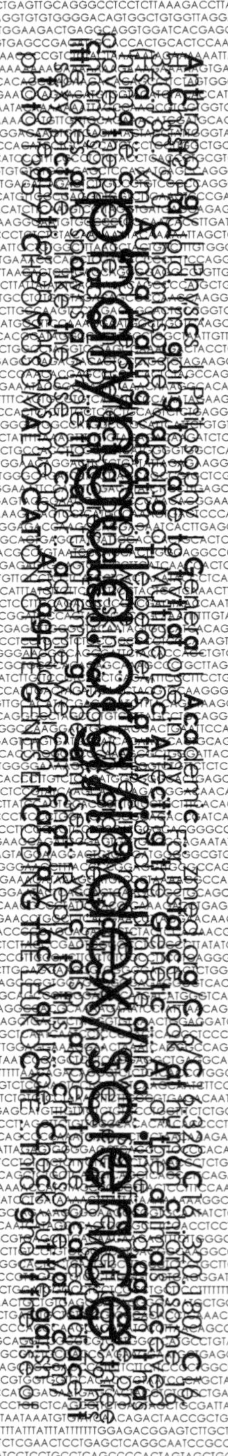

AND THIS DEFIES PROBABILITY

P

vii

One million five hundred and sixty-three thousand and forty-one

THAT UNIVERSAL AND PUBLIC MANUSCRIPT

D

One million six hundred and twelve thousand nine hundred and twenty-two

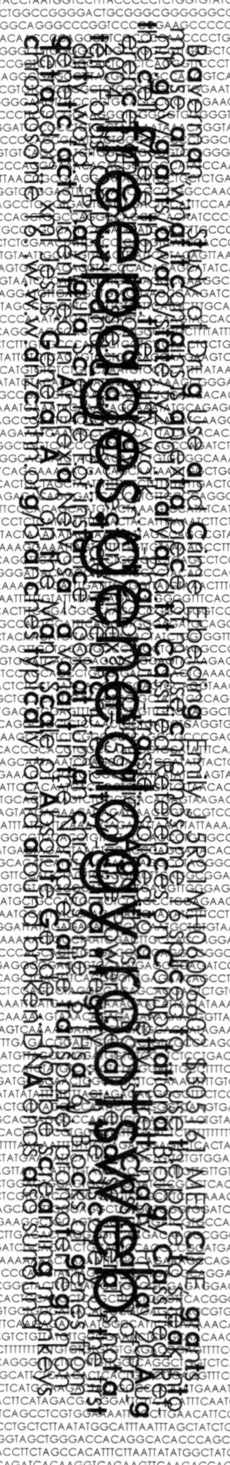

PLEASE EMAIL ME

F

viii

One million six hundred and sixty-three thousand four hundred and ninety-five

A STUDY SAYS
B

One million seven hundred and fourteen thousand seven hundred and seventy-nine

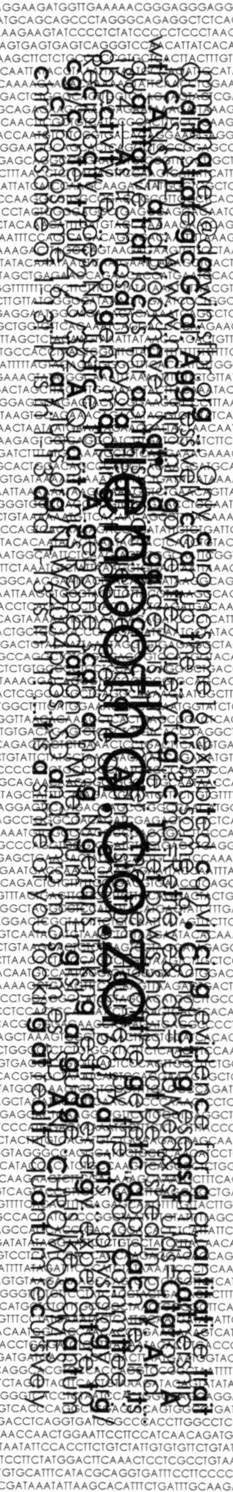

THE FATHER OF EUGENICS

L iv

en
wikipedia
data.org

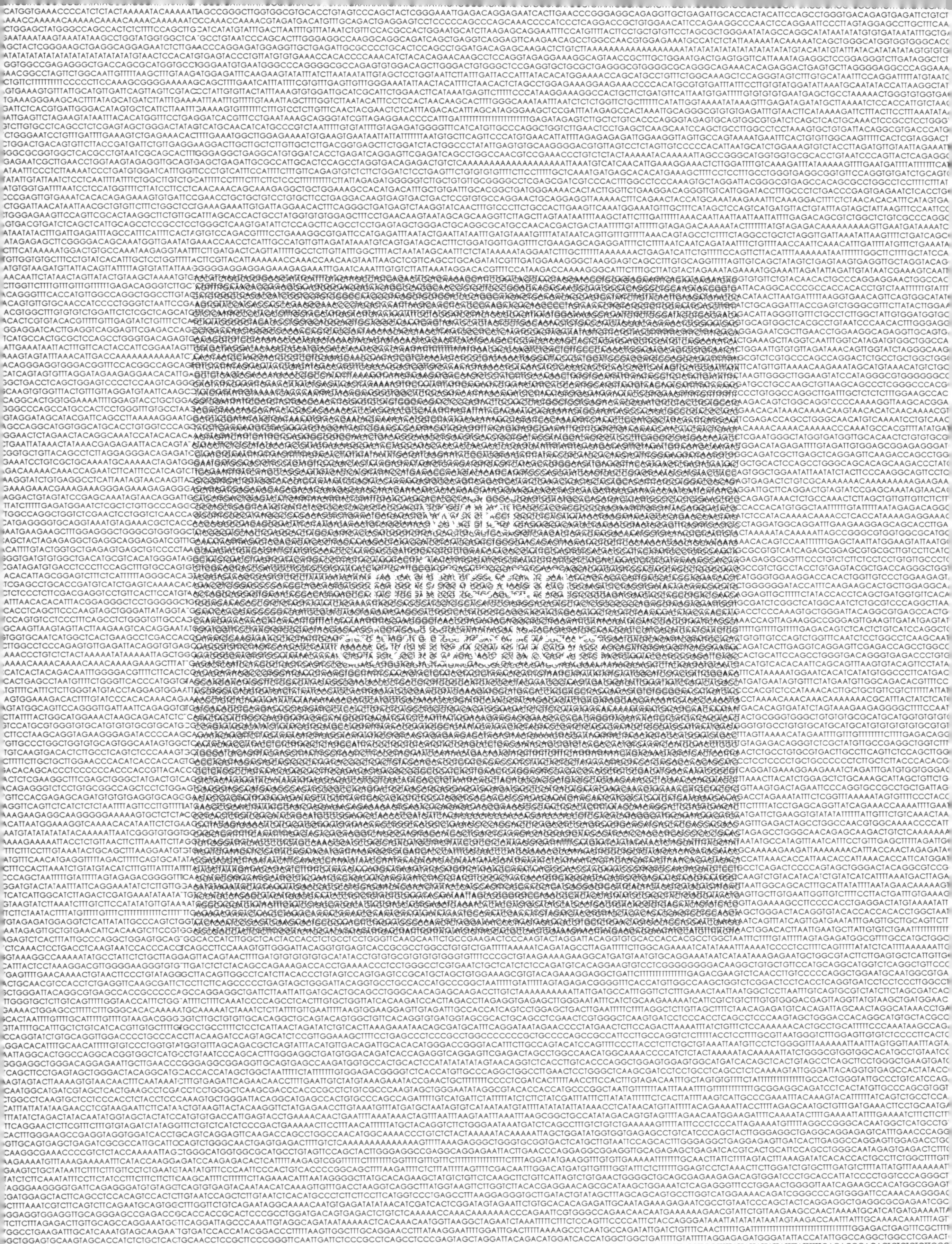

history
of
science
dotcom

g21
dees
timeline
timeline
1984
1991

[illegible]

fortune
city
dotcom

[illegible]

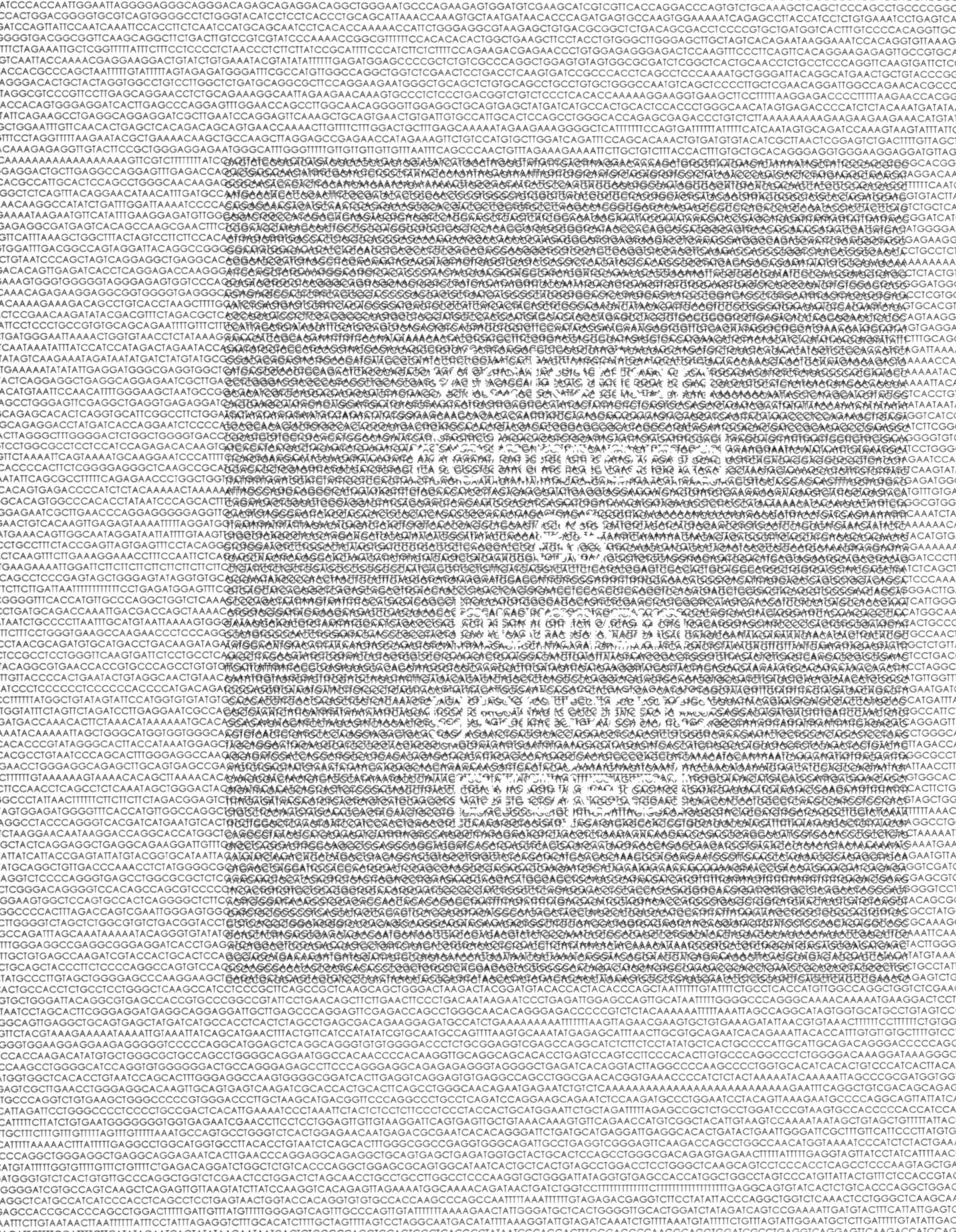

One million nine hundred and sixty-seven thousand five hundred and forty-five

SKIP THE DETAILS

F

Two million fifteen thousand nine hundred and eleven

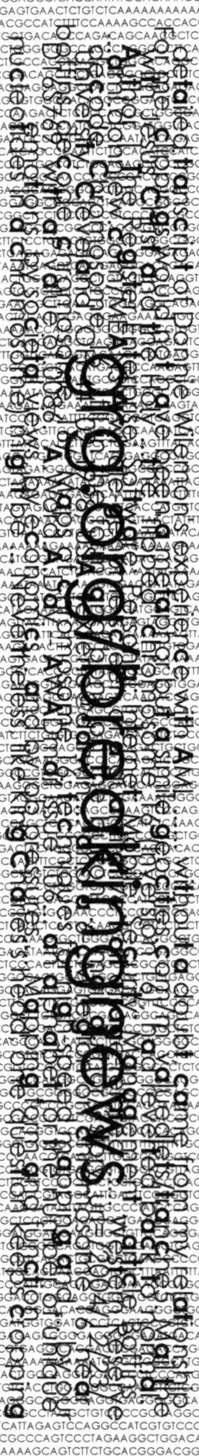

THE WORLD'S FIRST HUMAN BABY

G

xvi

Two million
sixty-five
thousand
one
hundred
and seven

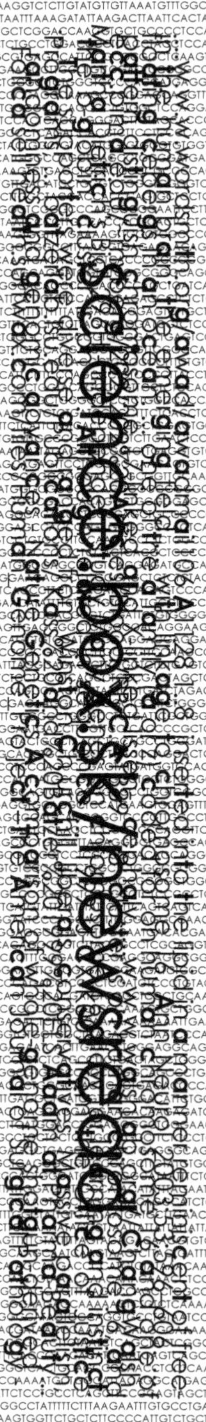

AN INFINITE NUMBER OF TIMELINES IN OUR UNIVERSE

S

Two million one hundred and fifteen thousand two hundred and fifty-nine

IT CAN BE SATISFYING TO SOLVE DIFFICULT PROBLEMS

N

xi

Two million
one
hundred
and sixty-six
thousand
three
hundred
and thirty-
one

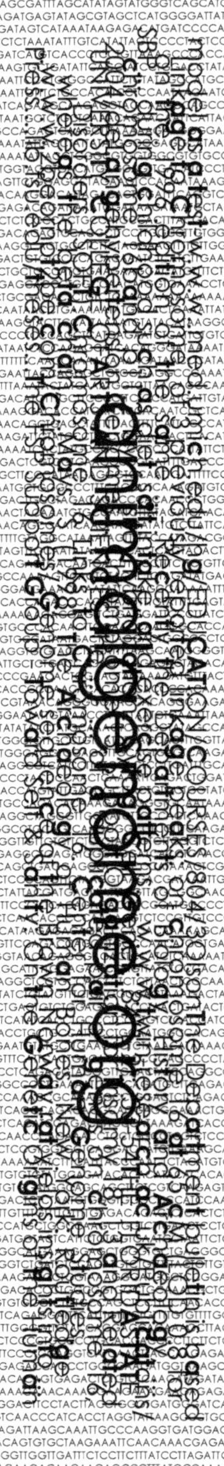

HORSE SHEEP CATTLE CHICKEN SWINE

A_{xy}

Two million two hundred and sixteen thousand two hundred and eighty-seven

G

vii

EVALUATE THE SCENARIO, SHORT VIGNETTES

Two million two hundred and sixty-four thousand six hundred and fifty-eight

THE PLACE OF THE MUSHROOM

C

Two million three hundred and thirteen thousand nine hundred and forty-four

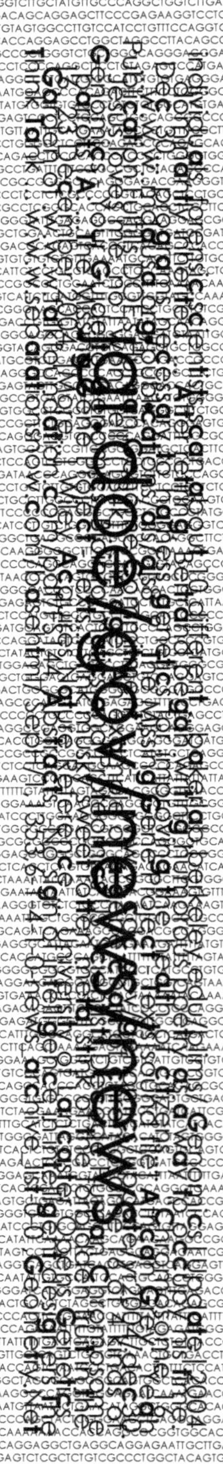

TERMITES FROM THE JUNGLES OF COSTA RICA

J

Two million three hundred and sixty-four thousand and fifty-three

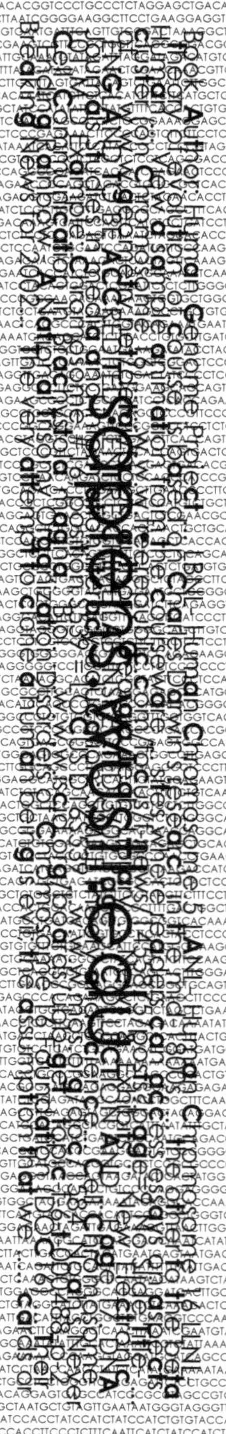

WHITE SPACES REPRESENT A GAP

S

Two million four hundred and fifteen thousand one hundred and seventeen

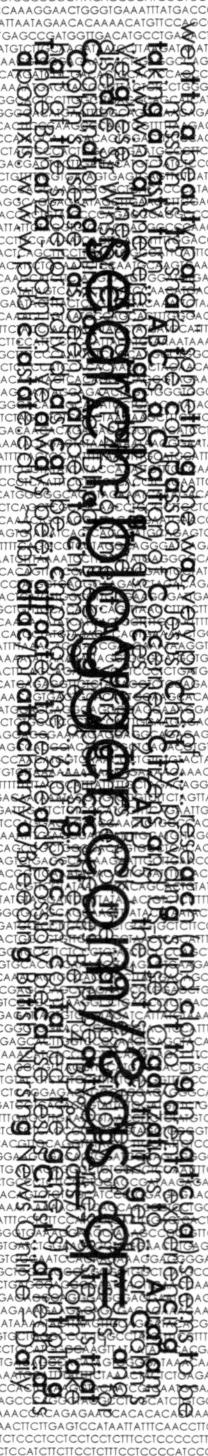

ACUTE AGITATION CONFUSION DISORIENTATION ANXIETY

S

Two million four hundred and six thousand two hundred and ninety-five

129.16.28.13/cipforum

nursing news & archives... not necessarily provide evidence but stressed that the injuries... that the chromosome
are variants of the MC1R gene, which is found on chromosome 16
16 – PRKCB1 seemed to have a connection with autism... Three judges sitting in London, Lord Justice Gage, Mr
Justice Gross and Mr... that the chromosome
composed of DNA (deoxyribonucleic acid)
Poetry is the universal language
COMPARISONS... But, in this study, you will learn that—even in DNA and chromosome counts—there is no
fever to the short arm of chromosome 16... www.banner-trader.com/tutorials/inflammation and Nephrotic
evidence of evolutionary descent. Evolutionary theory is a myth... www.pathlights.com/ce_encyclopedia/15sim
Syndrome: Fibrill. BUSTY ERICA... This conjures up a picture of DNA as a tangle of noodles swilling lazily around in

1

INVENTION → COMMERCIALIZATION

Two million five hundred and eighteen thousand and thirty

CONFIRM THE INTEGRITY OF MY REQUEST

Two million five hundred and sixty-nine thousand eight hundred and forty-three

WHAT'S THE COOLEST SITE ON THE WEB?

1x

Two million six hundred and twenty-two thousand and ninety-six

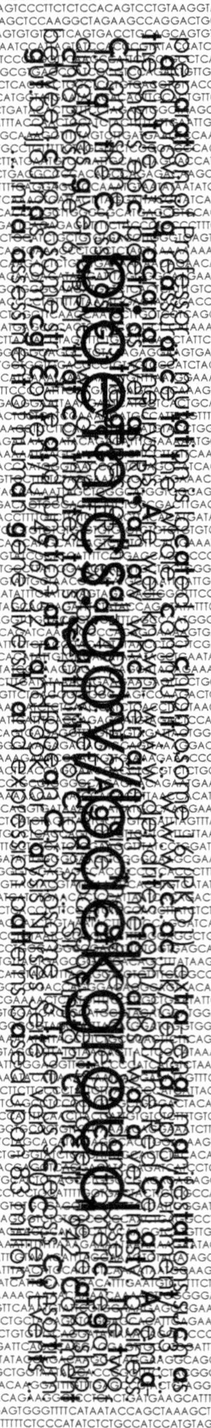

FOUNDATION FOR TRUTH IN REALITY

B

Two million six hundred and seventy-three thousand one hundred and ninety-six

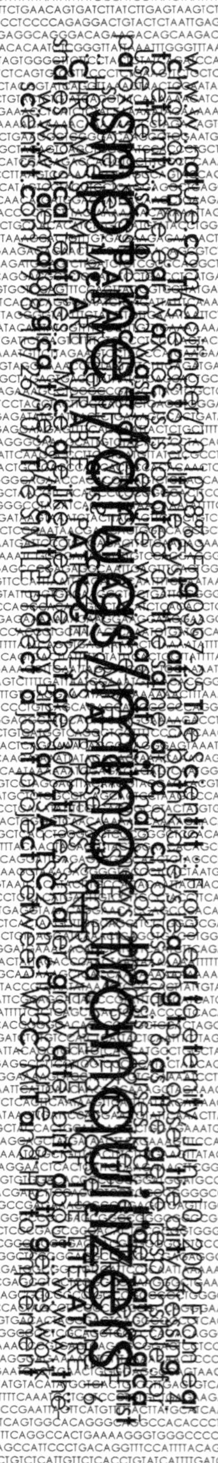

FROM A LARGE DISTRIBUTED UNIVERSITY SYSTEM

S_x

Two million seven hundred and twenty-four thousand one hundred and fifty-one

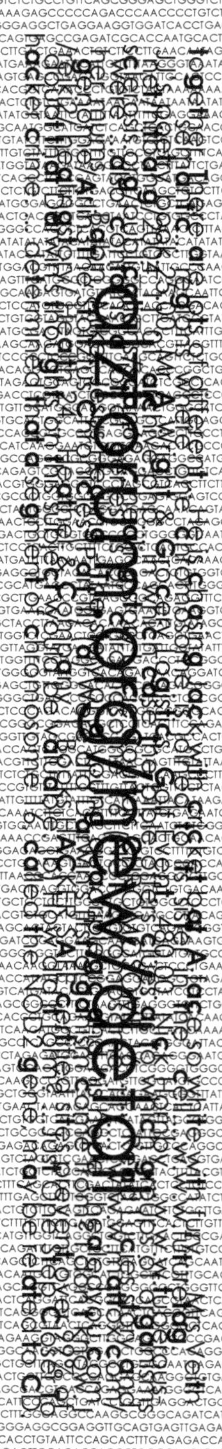

EXCITING OFF-MAINSTREAM PRESENTATIONS

A viii

Two million seven hundred and seventy-three thousand six hundred and sixty-four

OR HAVE WE ALREADY PASSED THAT THRESHOLD?

A xxii

Two million eight hundred and twenty-four thousand one hundred and seventy-three

SELECTED INTERESTING RELEVANT SENTENCES

A

Two million
eight
hundred
and
seventy-four
thousand
and thirteen

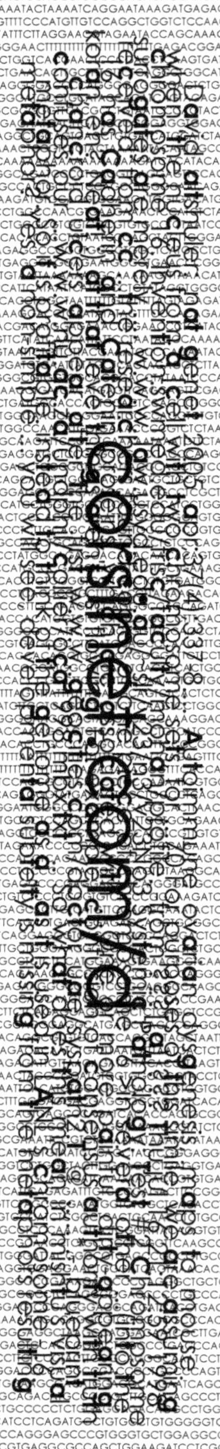

RED BOWS TWINNED BOWS COLOURED FRINGES DARK BANDS

C

xix

[illegible]

2
gcn
genomes
eukaryotes
homo
sapiens

Two million
nine
hundred
and ninety-
two
thousand
five
hundred
and fifty-six

WE MUST DISENTHRALL OURSELVES

L

Three million forty-five thousand two hundred and eighty-four

VIBRATORY MATRIX OF SOUND ITSELF

S xi

Three million
ninety-three
thousand
three
hundred
and
fourteen

Z

OA ~ ABLOOM ~ ABLUTION ~ OA

Three million
one
hundred
and fifty
thousand
and
seventeen

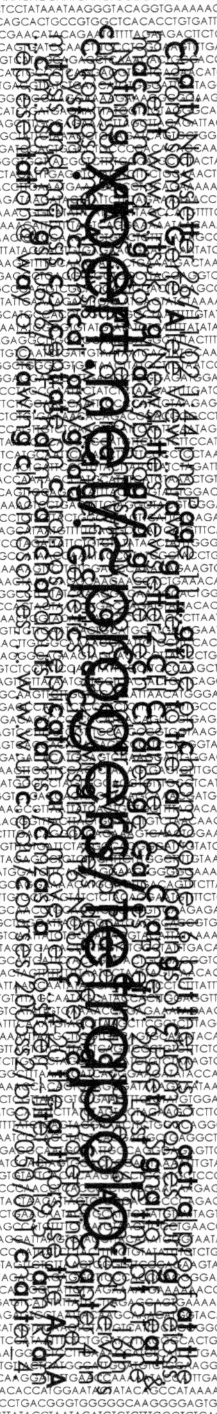

WE CAN RECOGNIZE IMPORTANT REVELATIONS

X

Three million two hundred and two thousand four hundred and seventy-seven

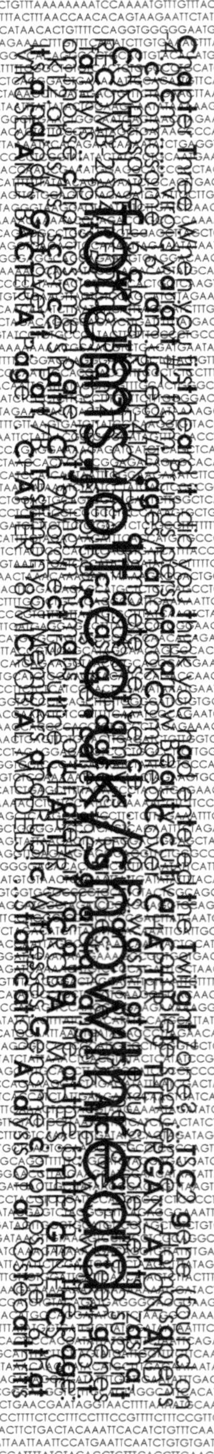

F

SILIBILI SUPERIOR GAMER

Three million two hundred and fifty-four thousand eight hundred and fourteen

CHILDREN IN HEAT DRIVE IN

T

Three million three hundred and seven thousand one hundred and thirty-two

AUDIO BITCH SLAPPED

R

Three million three hundred and fifty-eight thousand and seventy-four

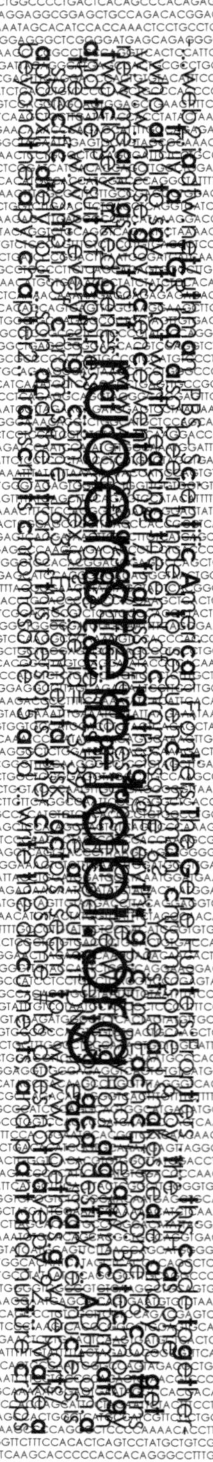

DO THE *RIGHT* THINGS AND EVENTUALLY YOU GET A BABY

R

Three million four hundred and eight thousand six hundred and sixty-one

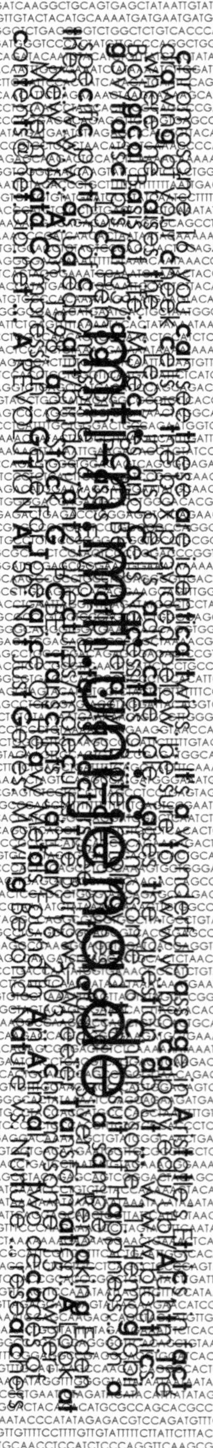

IMPLICATIONS

M xiii

Three million four hundred and sixty thousand one hundred and seventy-eight

NO PUBLISHING FEE
A
xxix

Three million five hundred and twelve thousand eight hundred and eighty-one

A WAY TO BREED TROUT THAT ARE BIGGER

J

Three million five hundred and sixty-four thousand eight hundred and eighty-one

THE NEED TO SCREEN OUT "JUNK SCIENCE"

O

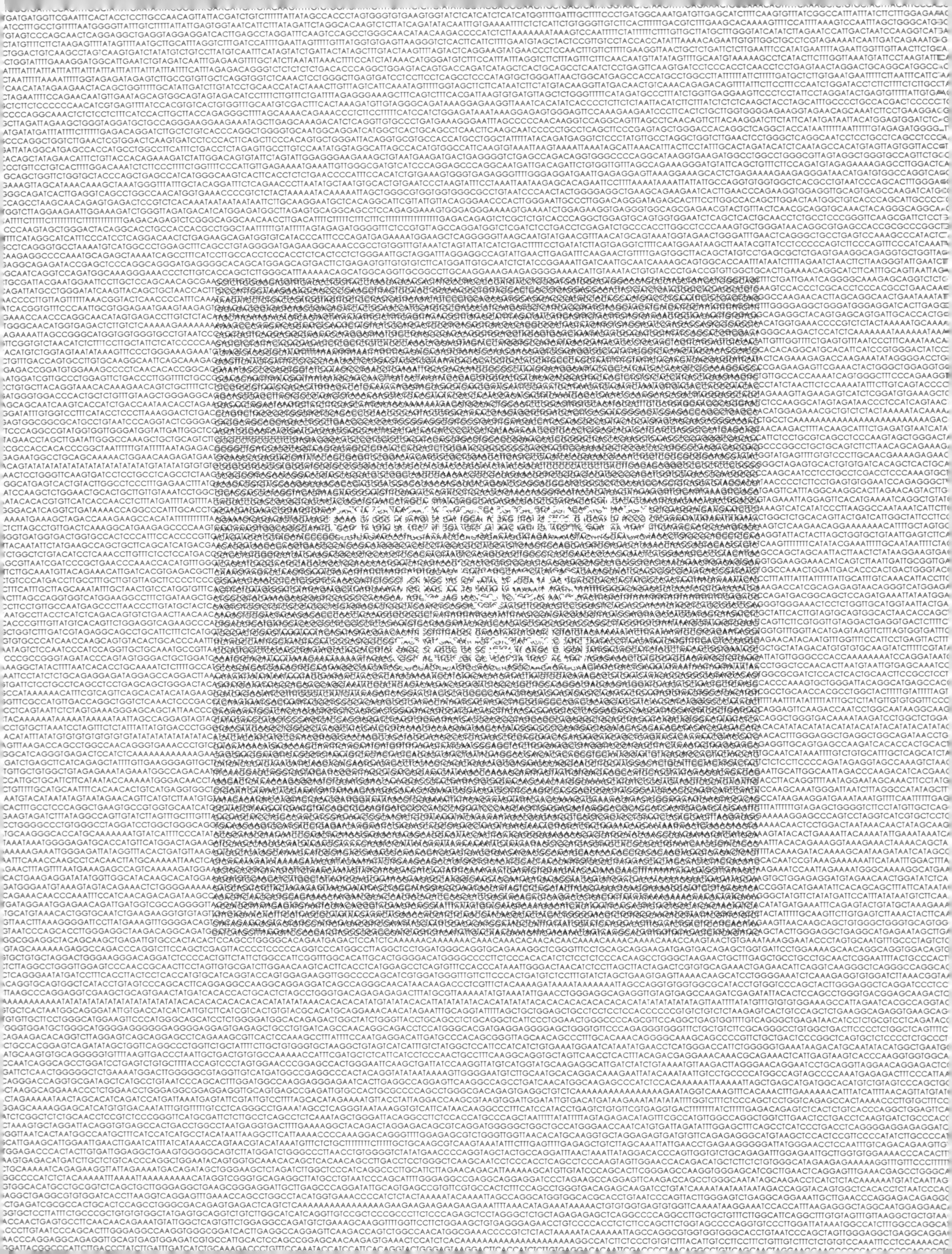

[illegible]

Three million six hundred and eighty-nine thousand nine hundred and ninety

WE ARE FORGING ON FULL STEAM AHEAD

Three million seven hundred and forty thousand six hundred and twenty-six

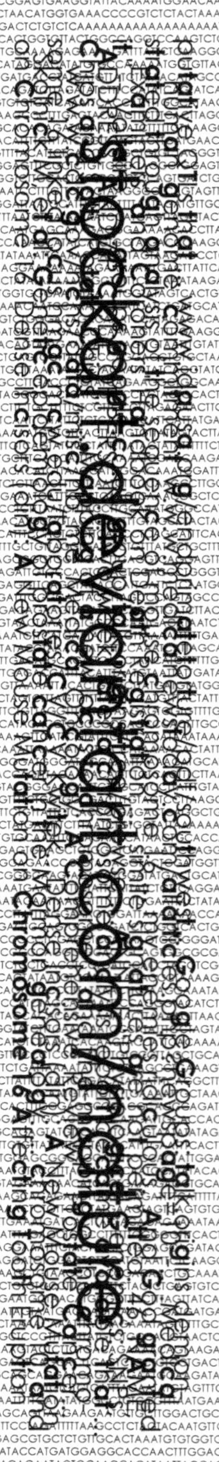

SHAMLESS SELF PROMAOTION

S

Three million seven hundred and forty thousand two hundred and twenty-two

NEW ABSENT HEALING PROGRAM AVAILABLE

C

Three million eight hundred and forty-seven thousand and forty

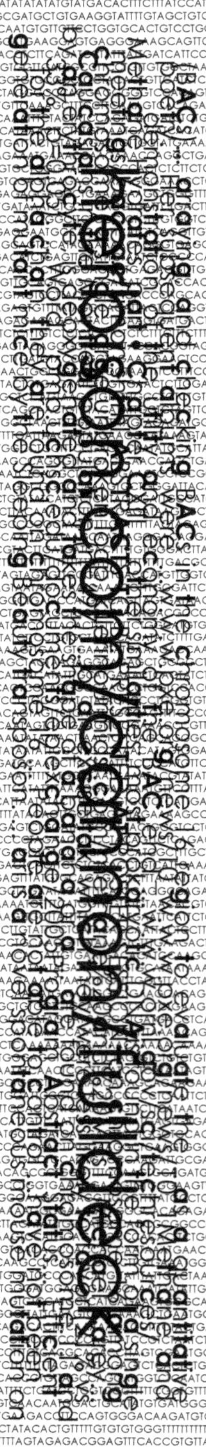

AND OTHER REALLY OFFENSIVE MATTER

H

Three million eight hundred and ninety-nine thousand two hundred and ten

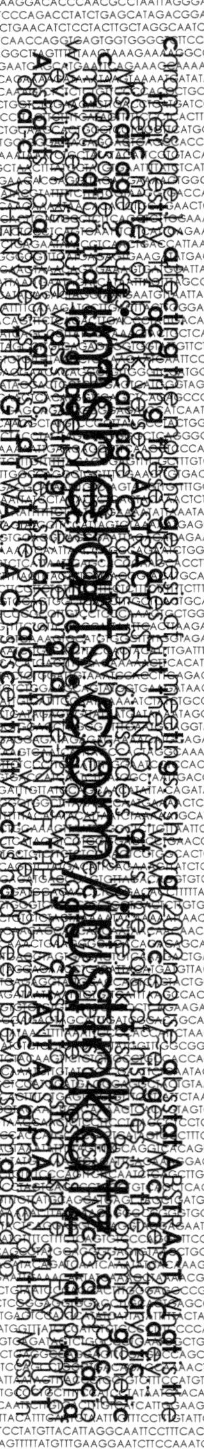

AND A COMMANDMENT OF CONTRADICTION

T

1x

Three million nine hundred and fifty-one thousand three hundred and fifty-four

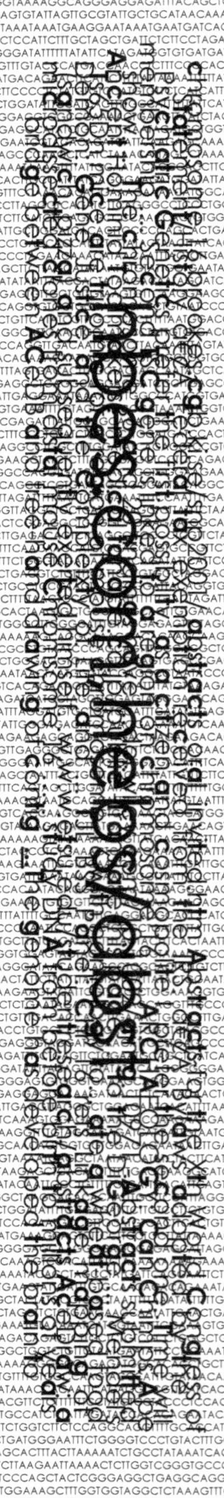

PERSONALITY AND HONEST COMMUNICATION

H

Four million four thousand one hundred and forty-three

THESE TWO *MICRO*-SETS DO NOT SHARE ANY BEAD TYPES

D

Four million
fifty-six
thousand
seven
hundred
and
seventeen

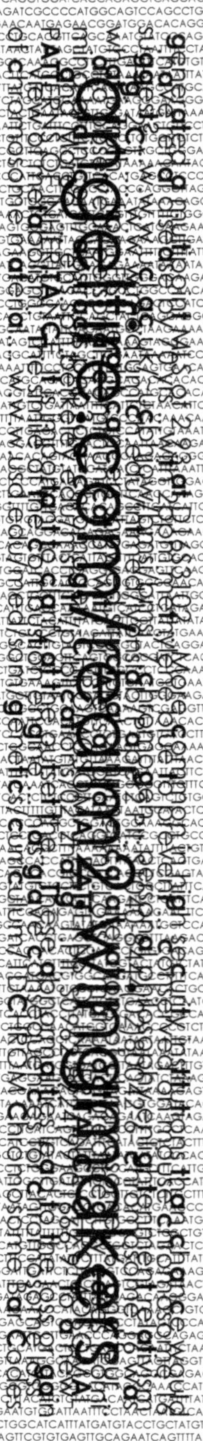

MEANING NO GLYPHS

A

xiv

Four million one hundred and nine thousand and eighty

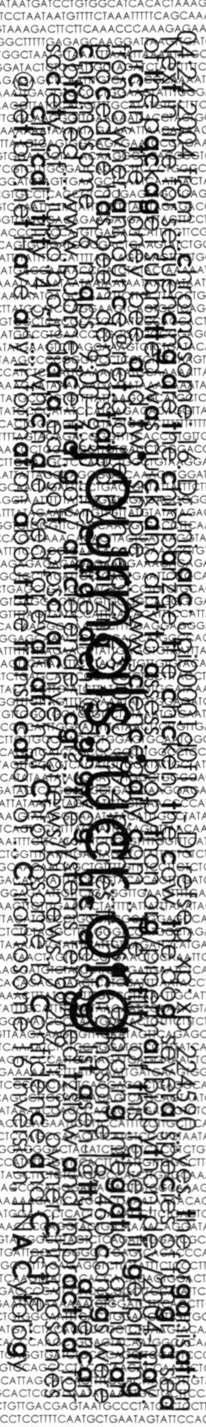

SYNERGY (HUMAN STUTTERING)

J

IV

Four million one hundred and sixty-one thousand seven hundred and forty-five

I USED TO PLAY WITH TINKER TOYS

M

Four million two hundred and fourteen thousand two hundred and twenty-seven

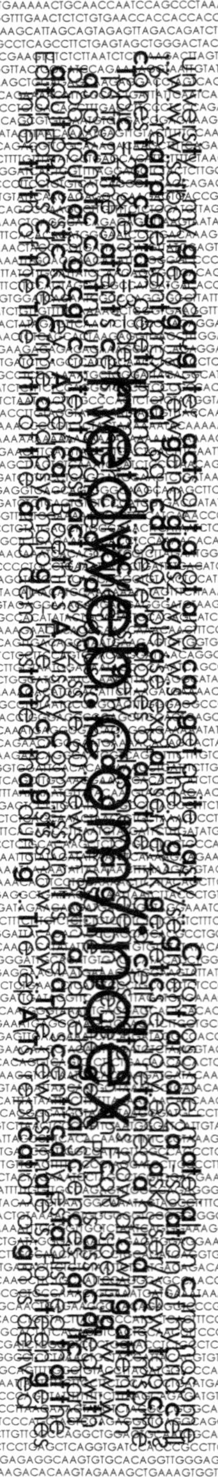

A PRECISELY DATABLE EVENT

Four million two hundred and fourteen thousand two hundred and twenty-seven

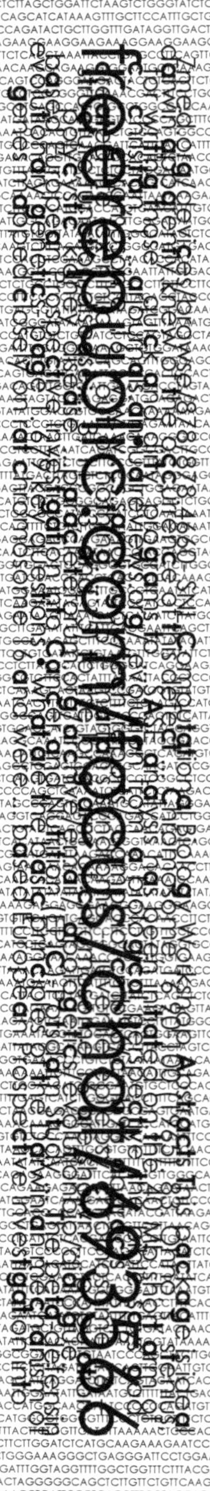

OVERALL OUR FRAGMENTS ARE LARGER

F

Four million two hundred and sixty-six thousand seven hundred and forty-five

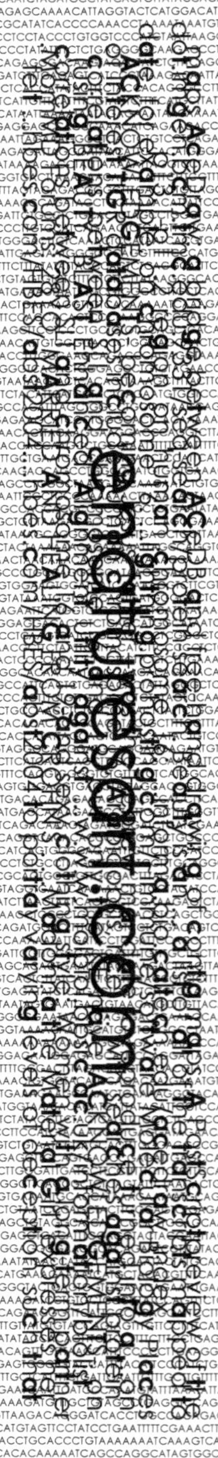

E

OUR *GIFT SHOP* HAS AN OUTSTANDING VARIETY OF *GIFTS*

Four million three hundred and eighteen thousand nine hundred and twenty-two

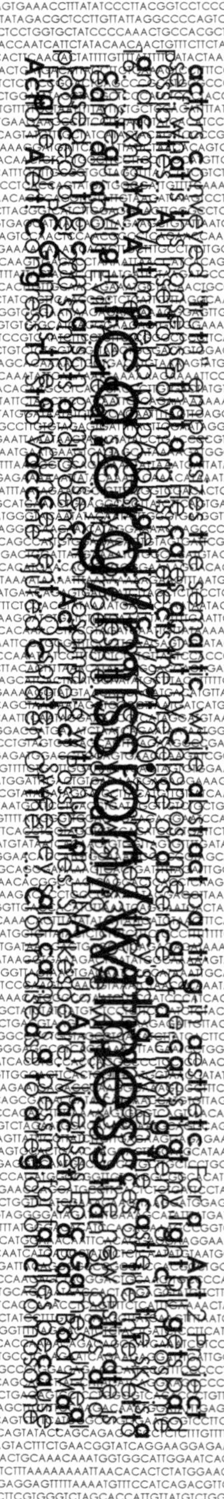

LEARN ABOUT YOUR METHOD OF THINKING ETHICALLY

R

Four million three hundred and seventy-one thousand and ninety-eight

EXCITING YEARS LIE AHEAD

Four million four hundred and twenty-three thousand and twenty-six

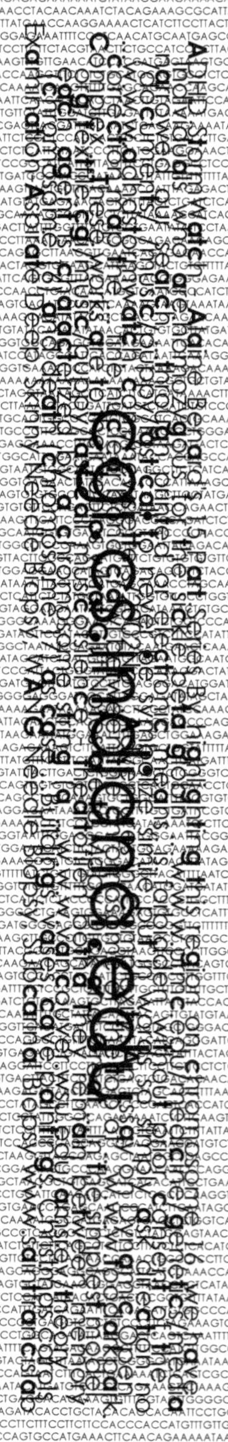

DISTILLED WISDOM AND SAGACITY

C_x

Four million four hundred and seventy-four thousand nine hundred and three

DESCRIBE THE INTRINSIC NATURE OF NATURE

viii

Four million five hundred and twenty-six thousand eight hundred and fifty-eight

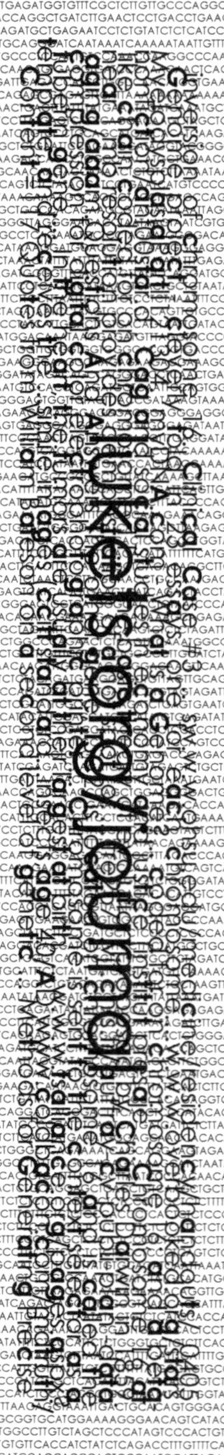

DEAR JOURNAL THINGS ARE FINE

L viii

Four million
seven
hundred
and
seventy-
seven
thousand
four
hundred
and
seventy-four

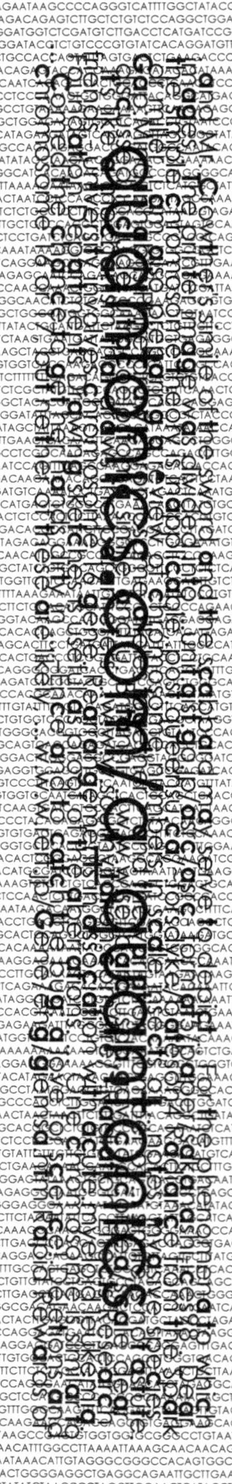

HOWEVER VISCOUS AND TENTATIVE THAT CHANGE MAY BE

Q

Four million six hundred and twenty-eight thousand five hundred and sixty-six

EVERYWHERE-BIVALENTLY-DISASSOCIATED

A

xix

Four million six hundred and eighty thousand five hundred and sixty-six

B

OF INTEREST TO FACILITIES WITH NON-HUMAN PRIMATES

Four million seven hundred and thirty-two thousand one hundred and fifty-seven

SPAM AND OFF-TOPIC CONTENT

Four million seven hundred and eighty-three thousand three hundred and sixty-five

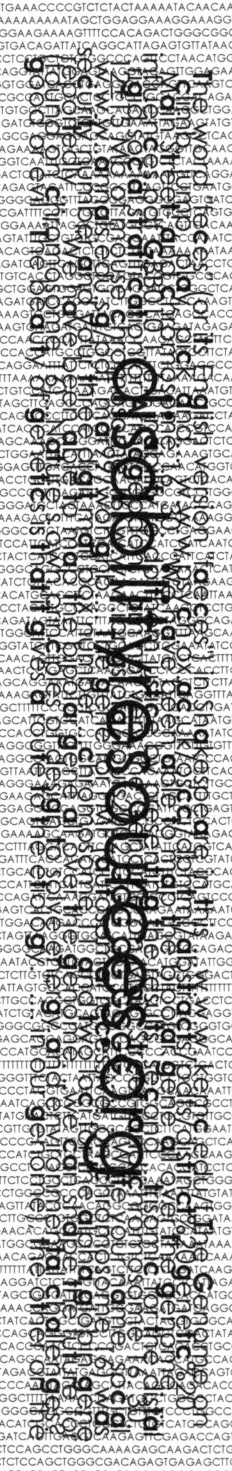

LIBRARIANS CONNECTIONS

D

Four million eight hundred and eighty-three thousand seven hundred and ninety-five

NOTHING BUT DEGREES OF AWARENESS

Four million nine hundred and thirty-three thousand six hundred and sixty-seven

P xiii

EVIDENCE FROM SIX-YEAR-OLD TWINS

Four million nine hundred and eighty-three thousand eight hundred and forty-three

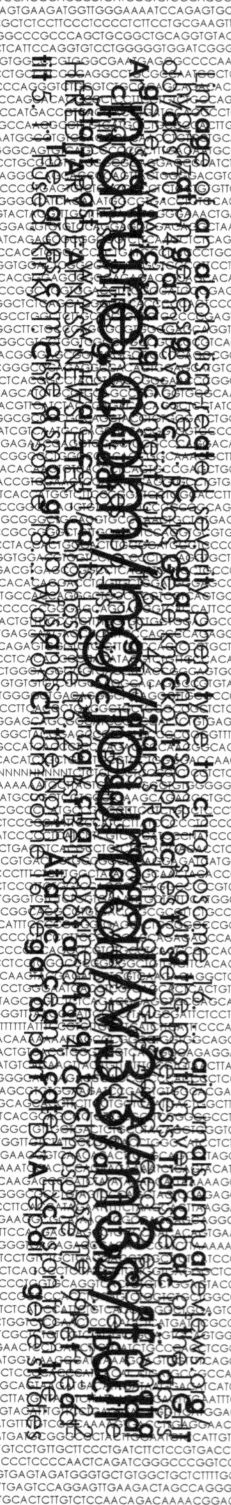

MASSIVE PARALLELISMS

N

Five million thirty-five thousand three hundred and thirty-four

VARICOSE VEINS
T
xy

Five million eighty-seven thousand six hundred and ninety-five

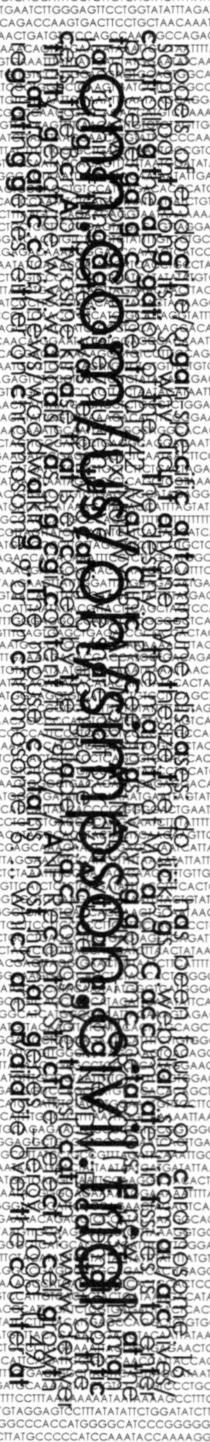

THE OFFICIAL REPORT ASSUMES NO RESPONSIBILITY

C

xv

Five million one hundred and thirty-nine thousand one hundred and eighty-seven

SCAN FOR LOCI PREDISPOSING TO SOCIAL PHOBIA

P

brown

rat

data

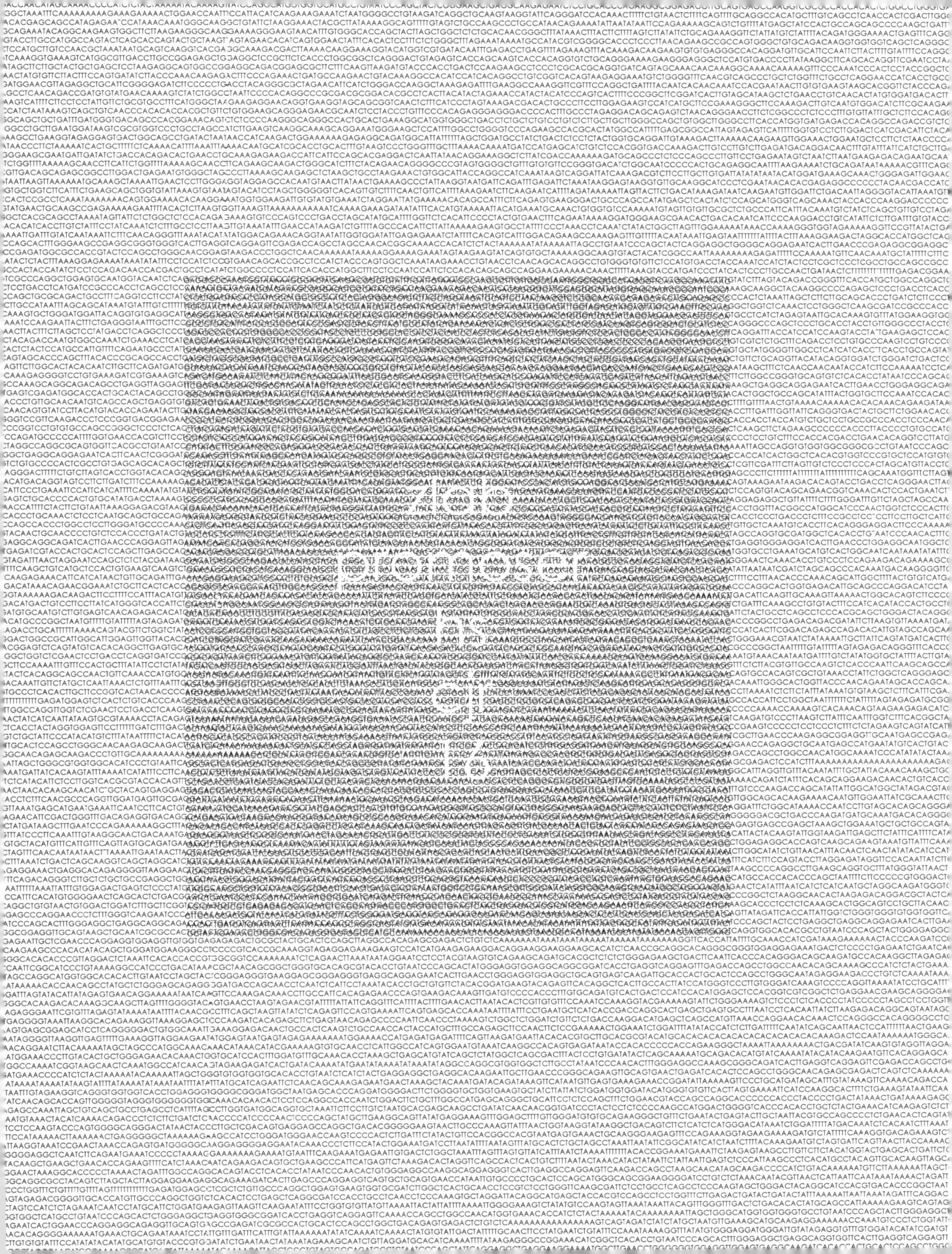

Five million two hundred and sixty thousand one hundred and seventy-eight

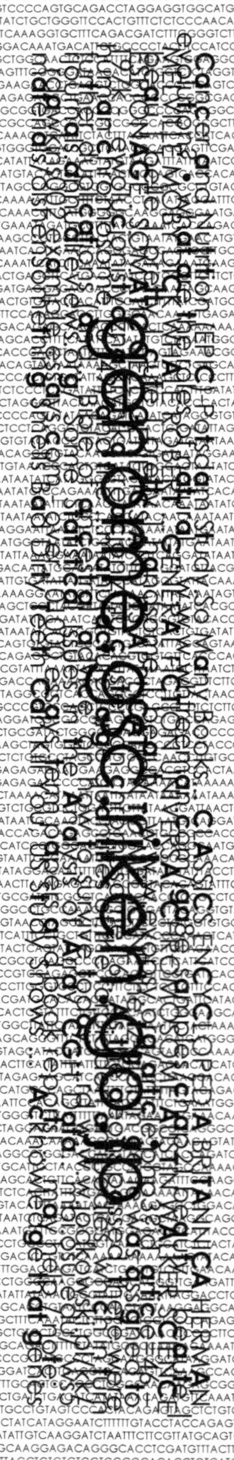

OFFICE OF PROTECTION FROM RES RISKS

Five million three hundred and ten thousand eight hundred and forty-two

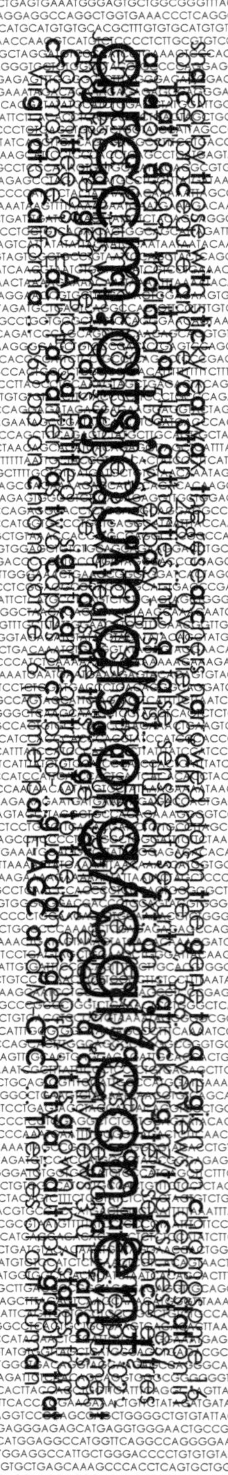

NINE TIMES IS NOT A GREAT MANY

A

Five million three hundred and sixty-one thousand three hundred and eighty-five

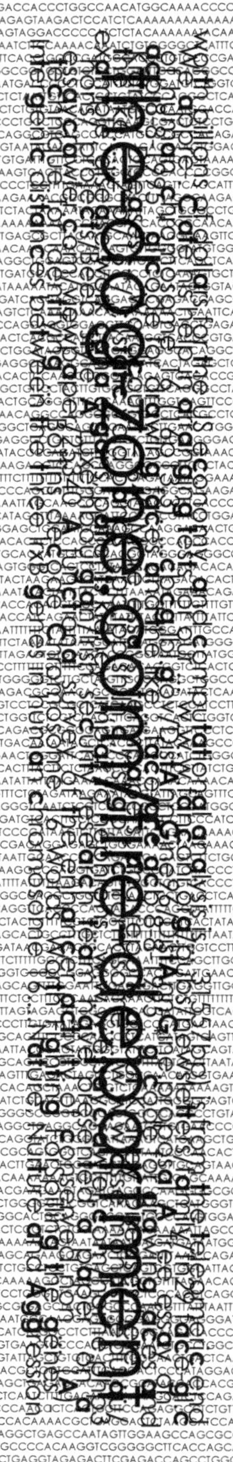

WEIRDER AND WILDER DOG SHAPES

Five million four hundred and eleven thousand seven hundred and two

DONATIONS CAN BE MADE ON LINE

L

Five million four hundred and sixty-four thousand six hundred and forty-three

S

xy

REAL NAME (HIDDEN)

Five million five hundred and sixteen thousand six hundred and ninety-eight

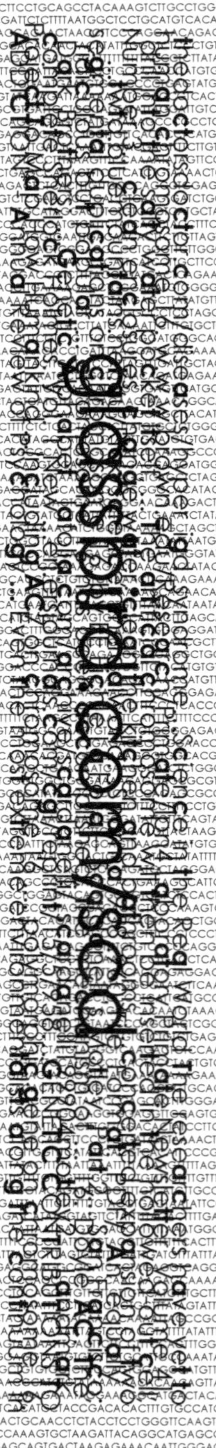

YES I AM PLAYING AROUND WITH THE LAYOUT OF THIS PAGE

G

xv

Five million five hundred and sixty-eight thousand one hundred and twelve

REALLY QUICKLY EVOLVING DYNOS

G

xiv

One hundred and eighty-three thousand three hundred and thirty

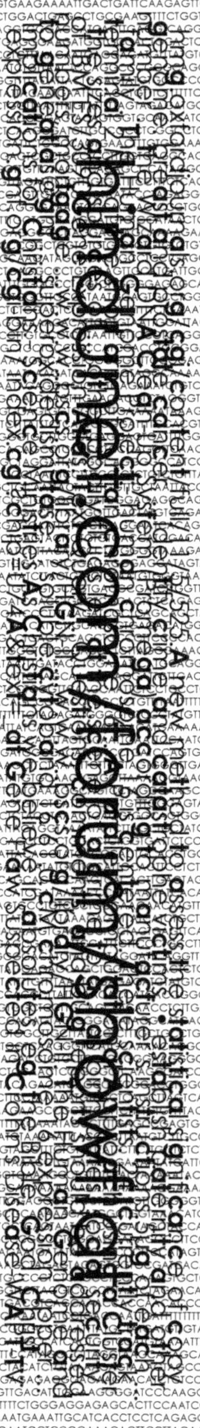

SELECTION CAN STIFLE VARIATION

H_{IV}

Five million six hundred and seventy-one thousand three hundred and seventy-one

W

WE NEED TO START ON TIME JUST IN CASE

Five million seven hundred and twenty-three thousand three hundred and six

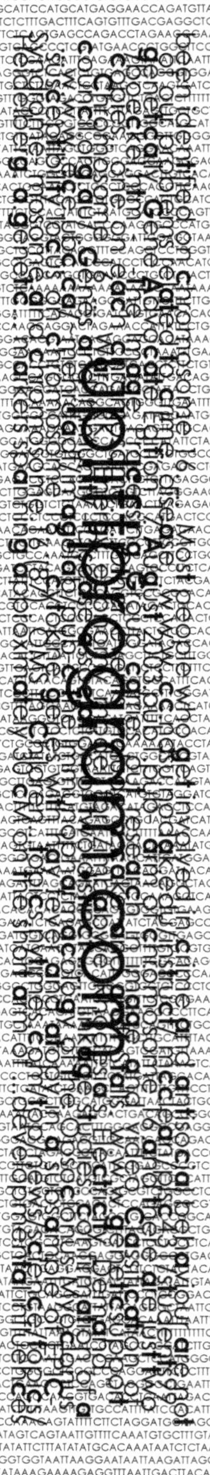

HEALTH AND WELLNESS NEWS

Five million seven hundred and twenty-four thousand four hundred and thirty-six

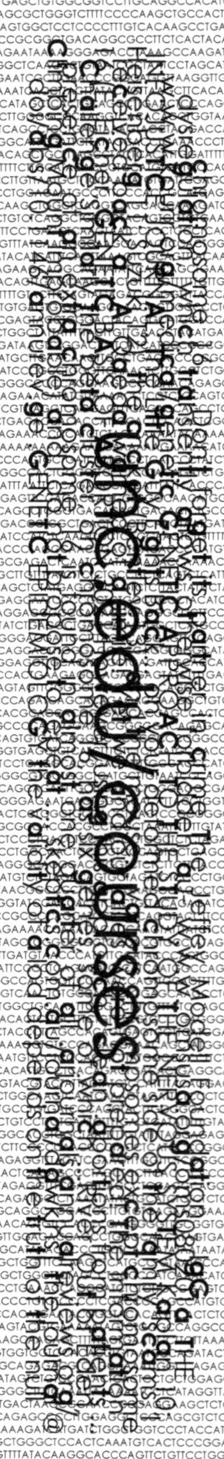

SKIP FIGURES 10, 11, 12, 13, 23
U

Five million eight hundred and twenty-five thousand five hundred and thirty

VIEW THE SIXTH GRADE CURRICULUM VIDEO

C

XVII

Five million eight hundred and seventy-six thousand one hundred and eighty-five

EXCEPTIONAL PENETRANCE

R vii

Five million nine hundred and twenty-seven thousand nine hundred and ninety-eight

READ THESE UNIQUE FAMILY STORIES

R

Five million nine hundred and seventy-nine thousand two hundred and nineteen

VIEW MY COMPLETE PROFILE

D

Six million
thirty-one
thousand
seven
hundred
and three

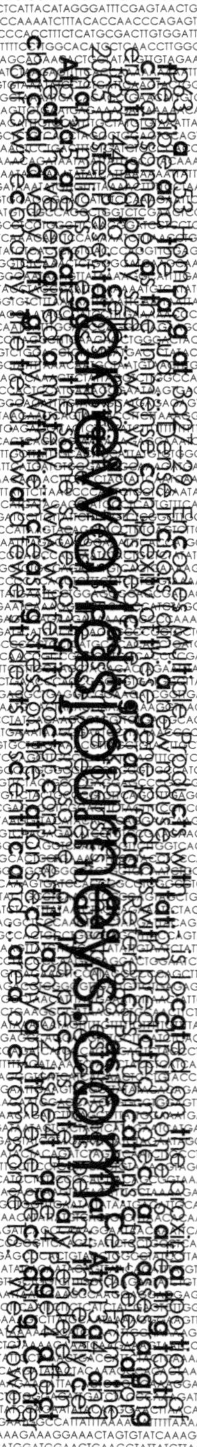

SHARE YOUR STORY

Six million eighty-three thousand one hundred and thirty-eight

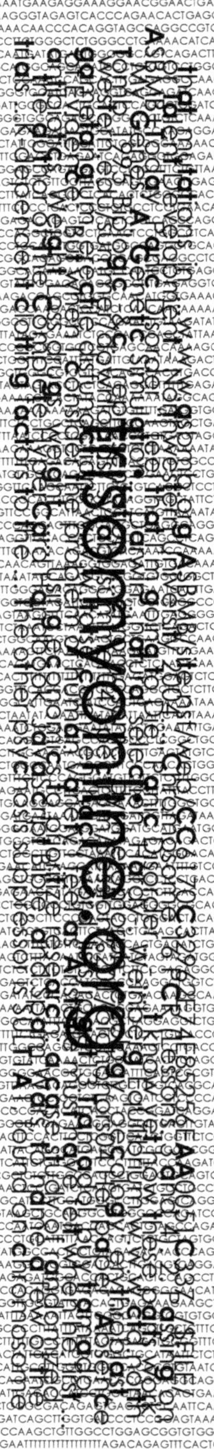

SITES OF OUR RAINBOW FAMILY

Six million
one
hundred
and thirty-
four
thousand
nine
hundred
and
seventy-
three

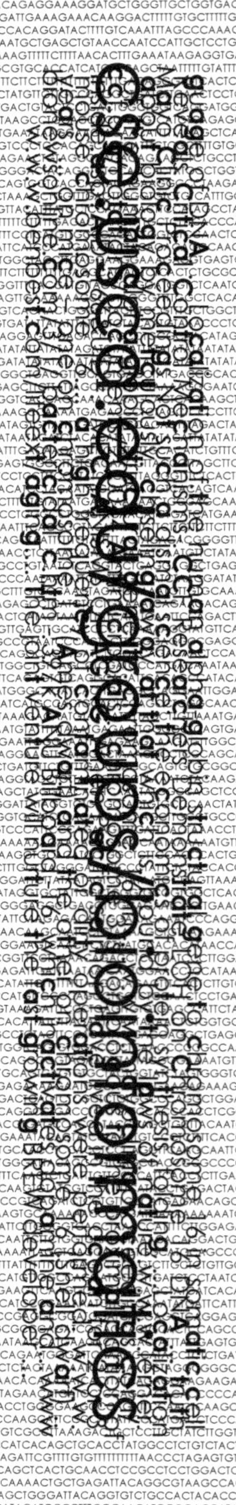

THIS FORMAT IS VALID BUT VERY SLOPPY

Six million one hundred and eighty-six thousand eight hundred and sixty-six

A MORE SUBTLE POINT (BUT A GOOD ONE)

P
VI

Six million two hundred and thirty-eight thousand six hundred and twenty

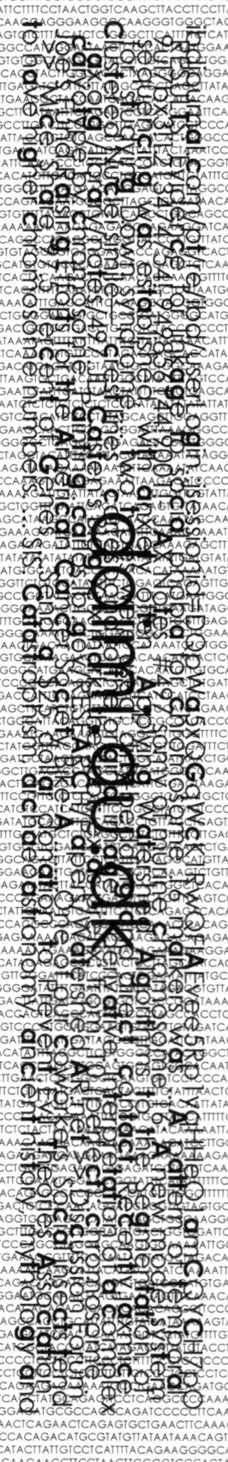

HEAT, 16? MICE, 11? BETWEEN, 9? FUNCTIONAL, 9?

D

Six million two hundred and ninety thousand four hundred and five

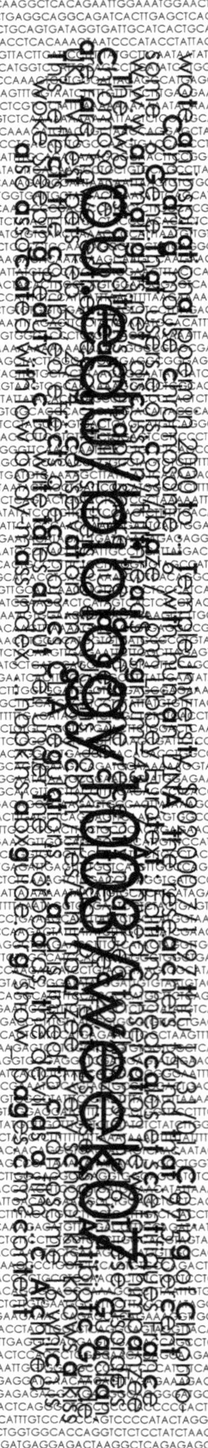

SUBJECT TO THE HONOUR CODE

Six million three hundred and forty-one thousand nine hundred and ninety-five

P

viii

SET YOUR COUNTRY TO VIEW

Six million three hundred and ninety-three thousand six hundred and nine

YOU CAN BE AN ADVOCATE

A

Six million four hundred and forty-five thousand seven hundred and ninety-two

YOU CAN CHANGE ALMOST EVERYTHING

T

Six million four hundred and ninety-eight thousand two hundred and eighty-nine

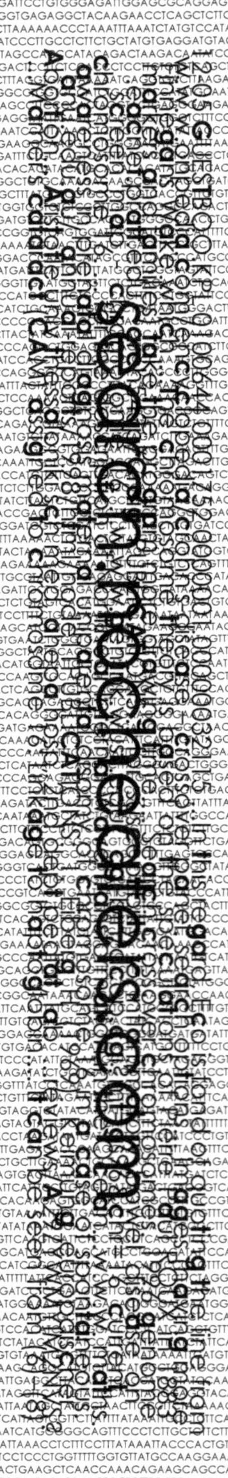

OUR WORK IS DESIGNED ONLY TO ASSIST

Six million five hundred and forty-nine thousand four hundred and twenty-six

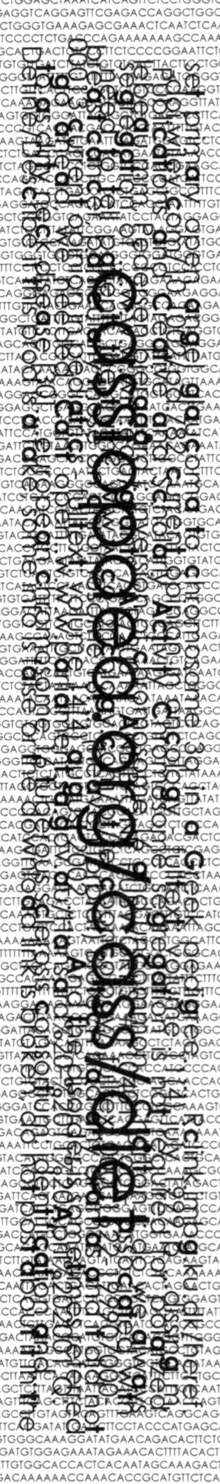

THE TRUTH ABOUT HYPERDIMENSIONAL BEINGS

C

Six million six hundred and one thousand one hundred and ninety-eight

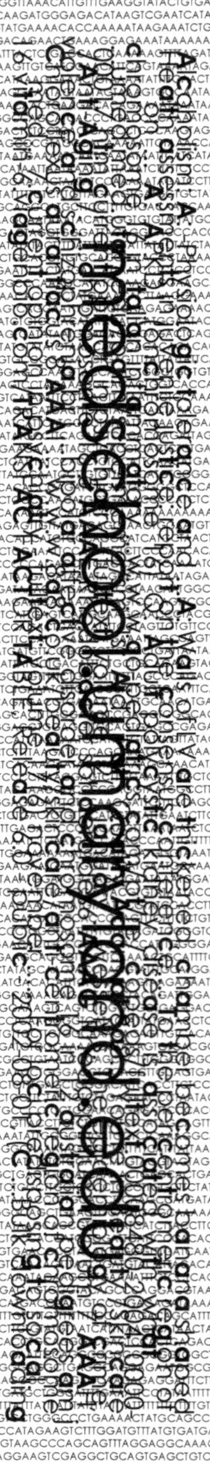

RESTRAINT DISINHIBITION HUNGER

M_{vj}

Six million six hundred and fifty-two thousand three hundred and twenty-six

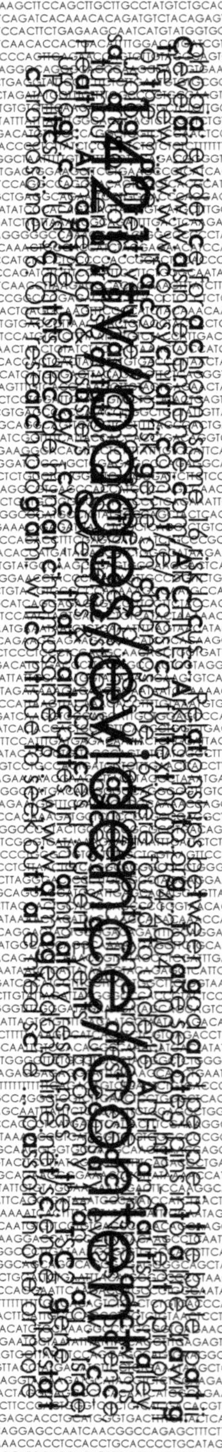

1

FROM ONE CONTINENT TO ANOTHER

Six million
seven
hundred
and four
thousand
and
nineteen

THINGS MUST BE KEPT IN PROPER PROPORTION

G

x1

Six million seven hundred and fifty-four thousand eight hundred and seventy

THE SCALE

S

vii

Six million
eight
hundred
and five
thousand
two
hundred
and sixty-six

AT PEACE AND SECURE

Six million eight hundred and fifty-eight thousand three hundred and twelve

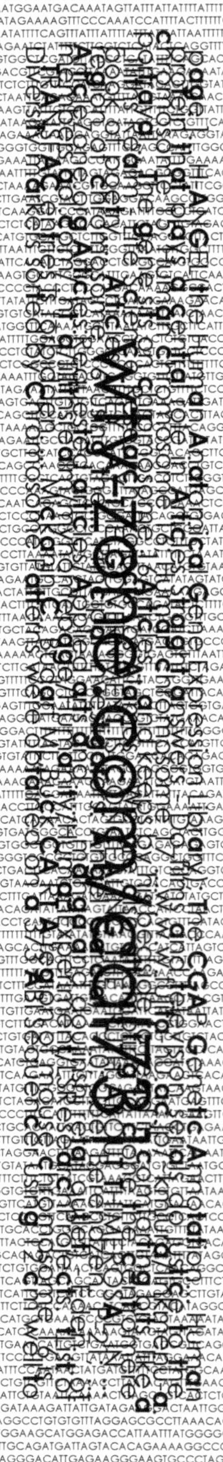

A SPECIAL SONG TO SING

Six million nine hundred and ten thousand one hundred and twelve

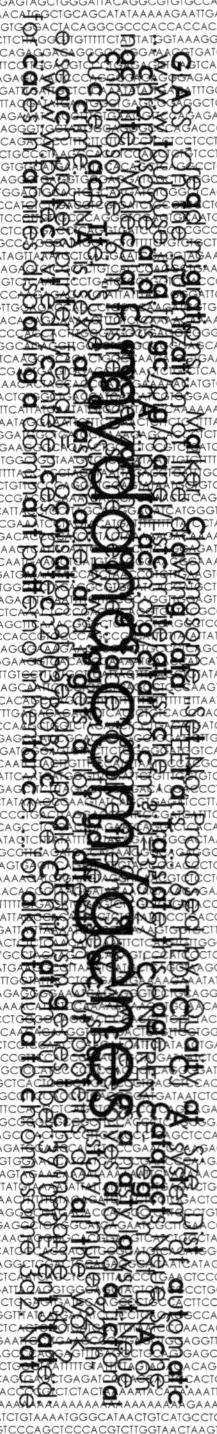

GOT A QUESTION? ASK HERE!

Six million
nine
hundred
and sixty-
two
thousand
three
hundred
and sixty-
one

WHAT IS A MOSAIC?
V

Seven million and thirteen thousand seven hundred and seven

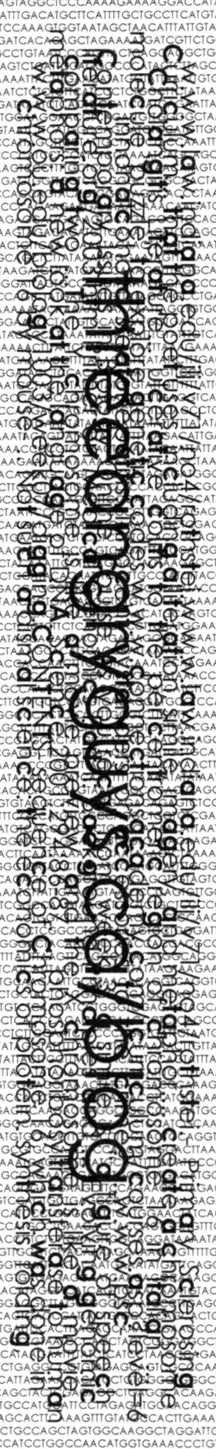

A GOOD 30 MINUTES OF FOREPLAY

Seven million and sixty-four thousand nine hundred and sixty-two

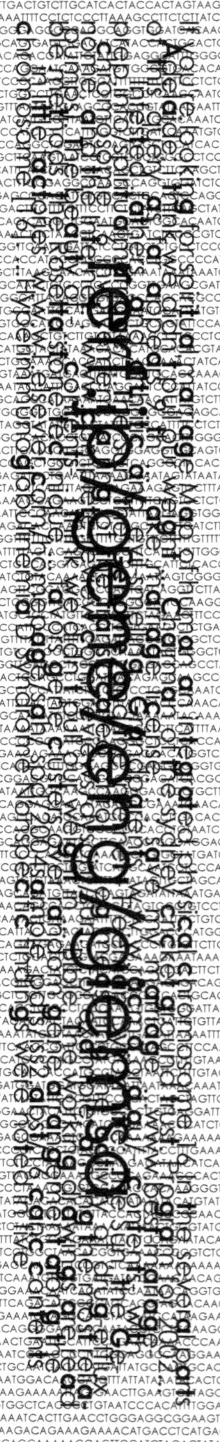

NOMENCLATURES FOR DESCRIBING ABERRATIONS

R_{1x}

Seven million one hundred and sixteen thousand four hundred and sixty-seven

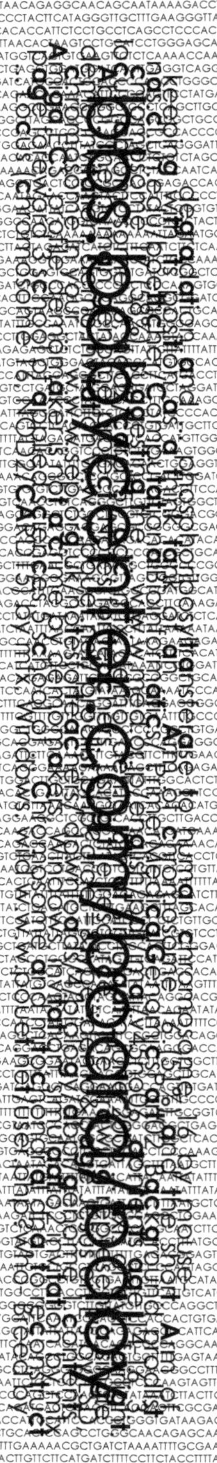

WHAT SHOULD I KNOW?

B

Seven million one hundred and sixty-eight thousand three hundred and thirty-eight

AN OPEN UNIVERSAL "MECHANISED" TEXT

G_{xi}

Seven million two hundred and twenty thousand four hundred and seventy-four

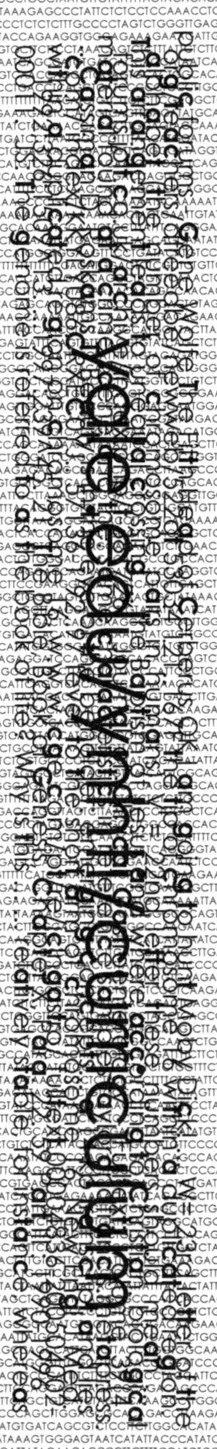

PREPARING AND DEFENDING A POINT OF VIEW

Y

Seven million two hundred and seventy-two thousand seven hundred and seventeen

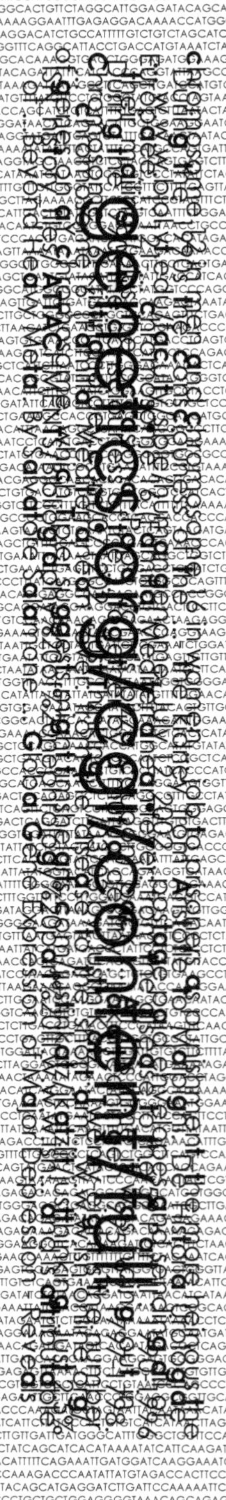

LINKAGE DISEQUILIBRIUM

G

VI

Seven
million three
hundred
and twenty-
four
thousand
four
hundred
and sixty-
nine

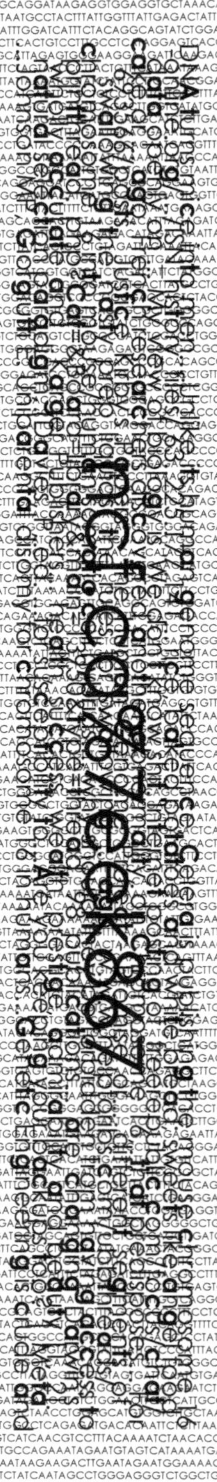

WE ARE IN OUR NATURE TREMENDOUSLY FLAWED

N

Seven
million three
hundred
and
seventy-six
thousand
seven
hundred
and nine

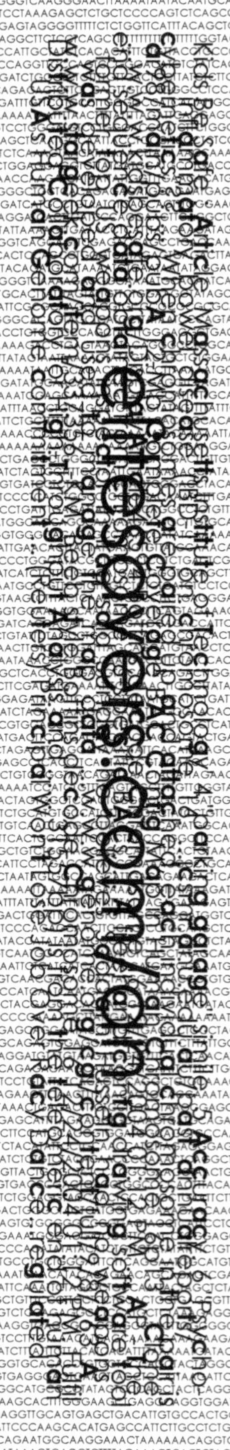

FREE AQUARIUM SCREENSAVER

E

Seven million four hundred and twenty-eight thousand six hundred and eighty-eight

N
STANDARDS OPINIONS EDITORIALS

Seven million four hundred and eighty thousand five hundred and sixty-four

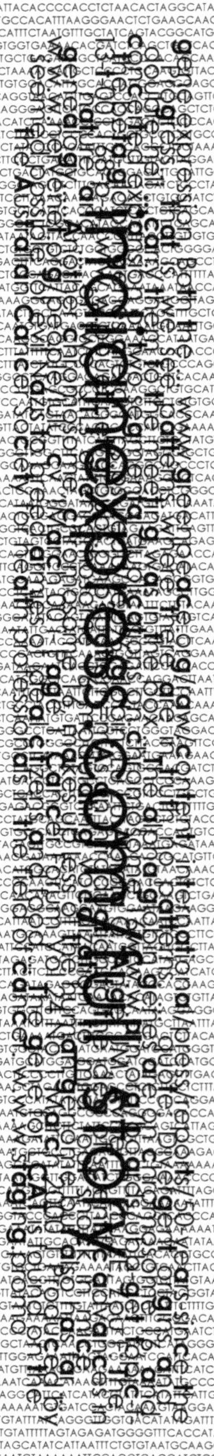

ALL HEAD IN THE SAME DIRECTION

[illegible]

sanger
data
dump

[illegible]

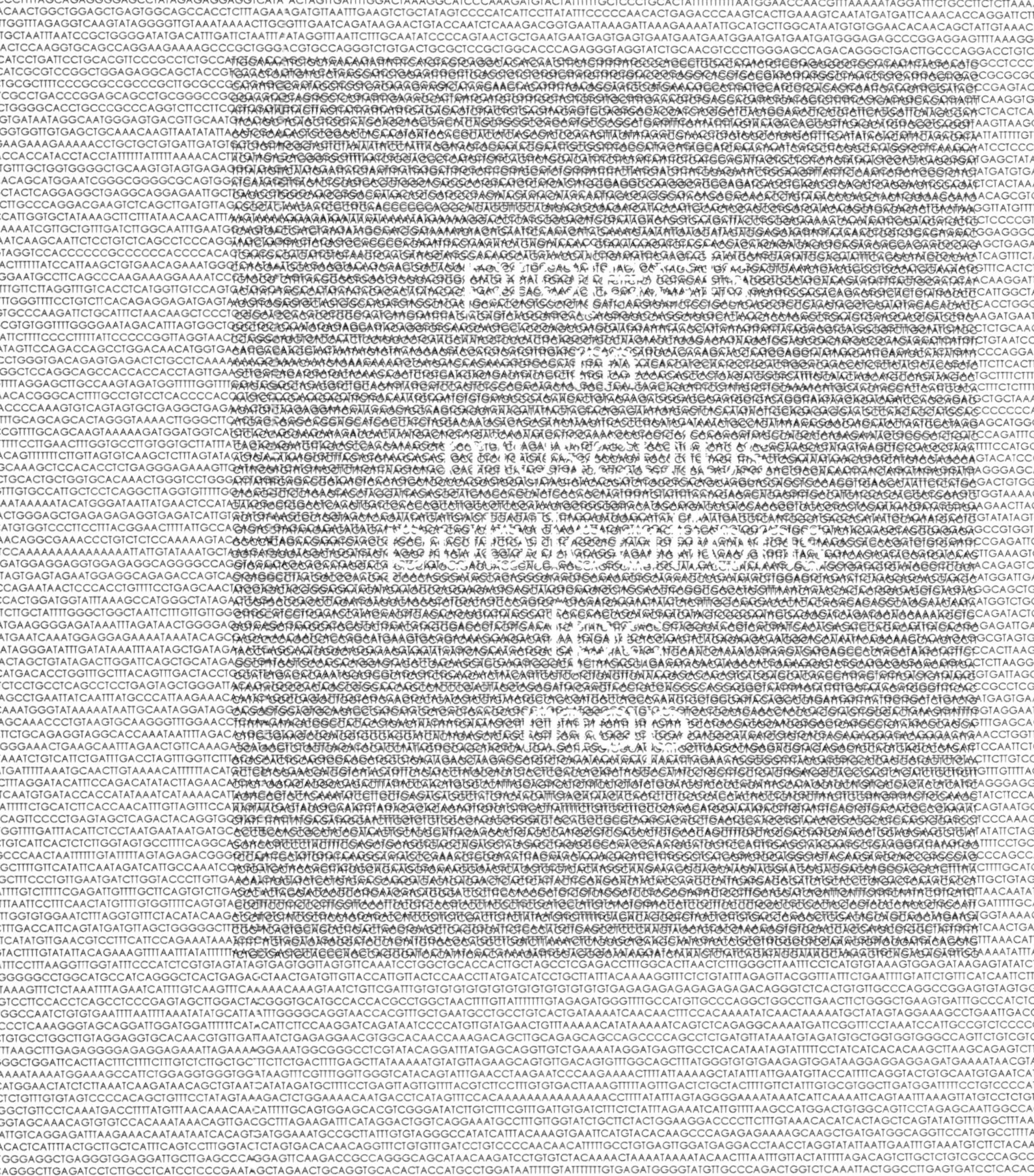

Seven million six hundred and four thousand six hundred and forty-three

ENTHUSIASTIC ABOUT LEARNING

E

vii

Seven million six hundred and fifty-seven thousand six hundred and twenty-three

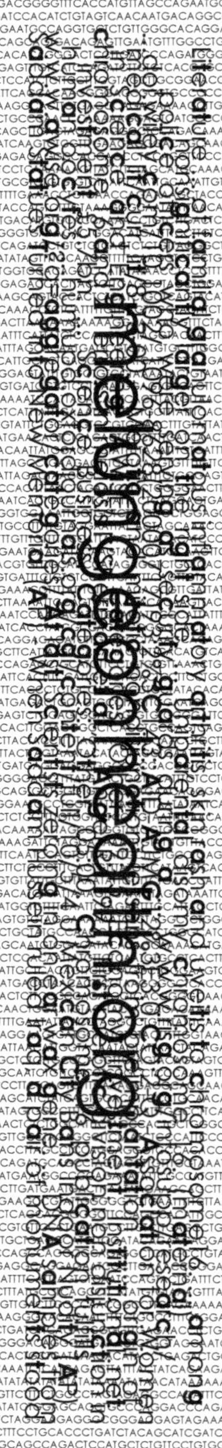

HELLO! I AM NANCY SPARKS MORRISON

M

vii

Seven million seven hundred and nine thousand five hundred and fifty-eight

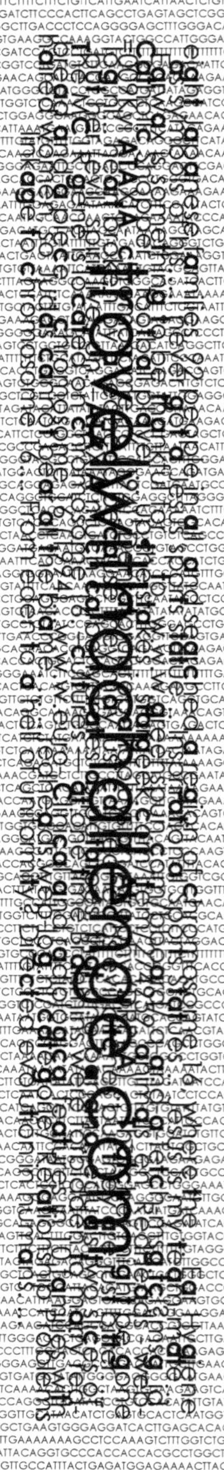

WELL-ESTABLISHED NEWLY-MINTED

T

xii

Seven million seven hundred and sixty thousand five hundred and ninety-six

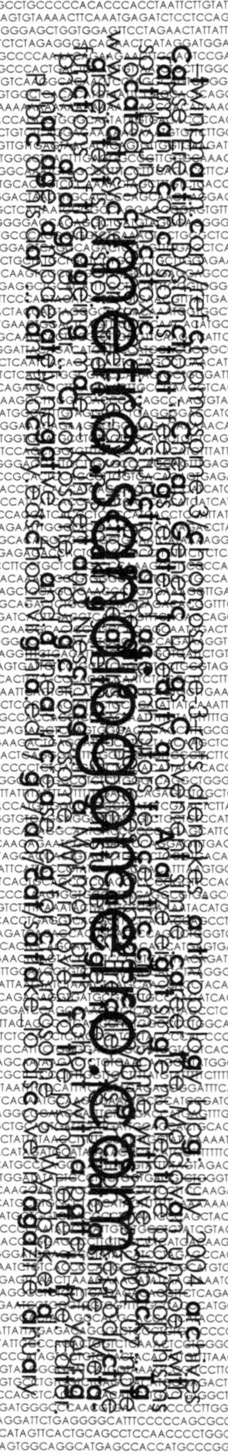

LASER-BASED COMBAT TRAINING SYSTEM

M fx

Seven million eight hundred and eleven thousand seven hundred and sixty-two

AUTOMATED ELDER ABUSE LITERATURE SEARCH

C

v1

Seven million eight hundred and sixty-two thousand seven hundred and sixty-eight

WRITTEN OUT IN ITS ENTIRETY

Seven
million nine
hundred
and thirteen
thousand
nine
hundred
and fifty-five

ONE FORGETS THE TIGER HEART

K

Seven
million nine
hundred
and sixty-
five
thousand
four
hundred
and eighty

NO ONE CAN TELL ME WHAT IT ALL MEANS

E

xii

Eight million sixteen thousand six hundred and forty-two

CURRENT EVENTS AND CONSPIRACIES

A$_{XX}$

Eight million sixty-eight thousand two hundred and fifty-eight

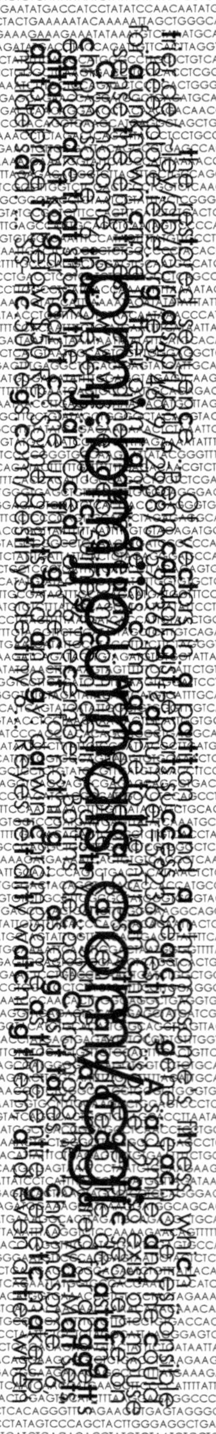

ERECTILE DISFUNCTION IN RATS

Eight million one hundred and twenty thousand four hundred and eighty-three

JOE DOES HAVE A WAY WITH PARODY

A xvi

Eight million one hundred and seventy-eight thousand two hundred and sixty-seven

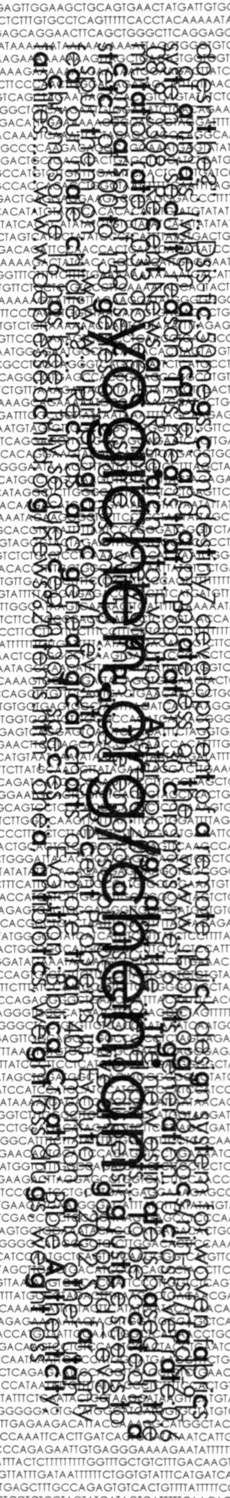

MANY BOOKS ON BOTH SIDES FOR AND AGAINST

Y

Eight million one hundred and seventy-eight thousand two hundred and sixty-seven

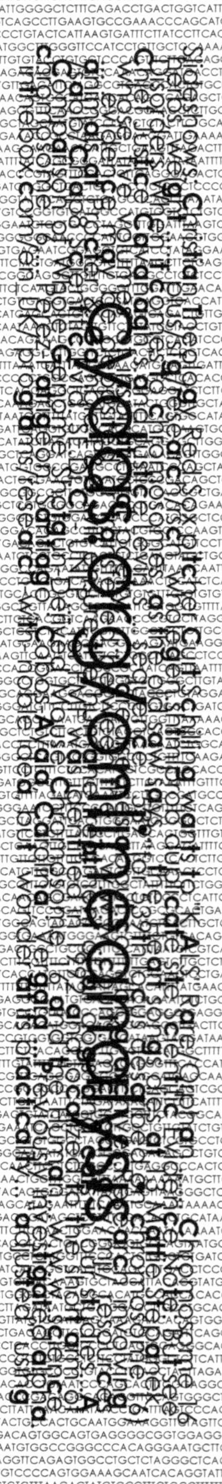

BETWEEN PAIRS OF REARRANGEMENTS

C xxiii

Eight million two hundred and seventy-six thousand four hundred and forty-two

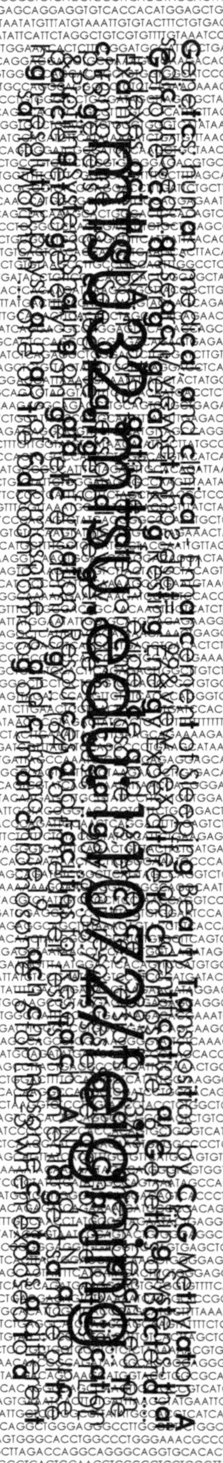

ON ASYLUM AVENUE

M

XIV

Eight million three hundred and twenty-eight thousand three hundred and sixteen

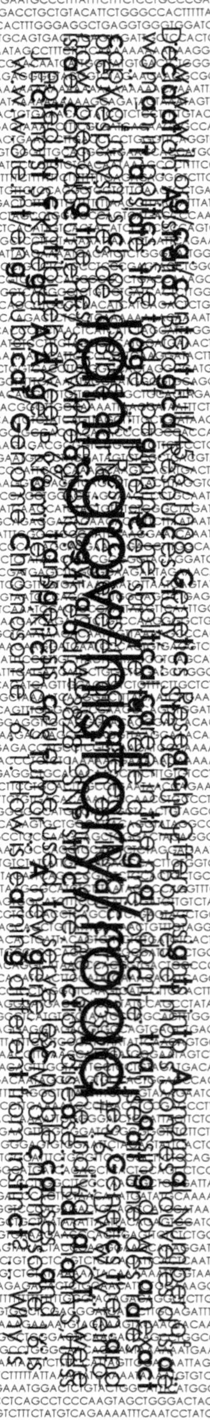

FOLLOW THE BALL OF RISING FIRE

L

Eight million three hundred and eighty thousand seven hundred and seventy-five

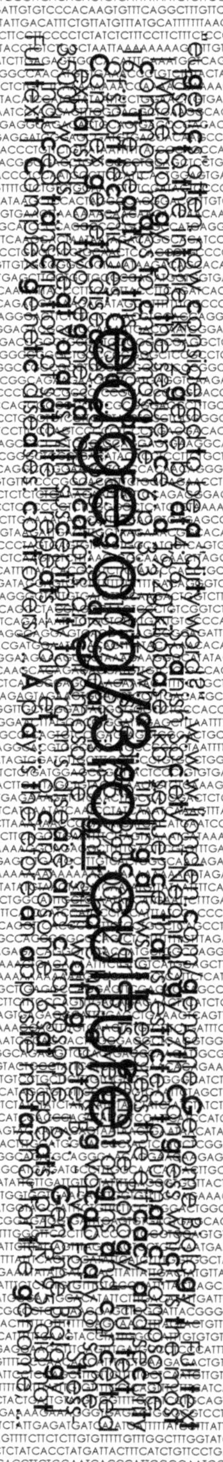

JUST THE OPPOSITE
E

Eight million four hundred and thirty-three thousand four hundred and seventy

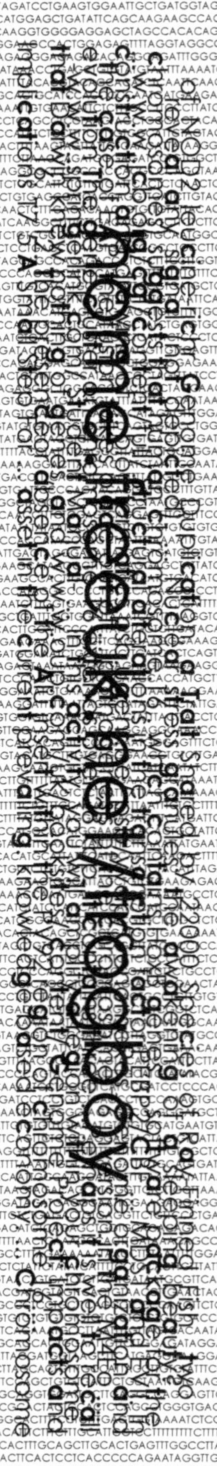

A RANDOM CONVERSATION

H

vii

Eight million four hundred and eighty-five thousand seven hundred and twenty-four

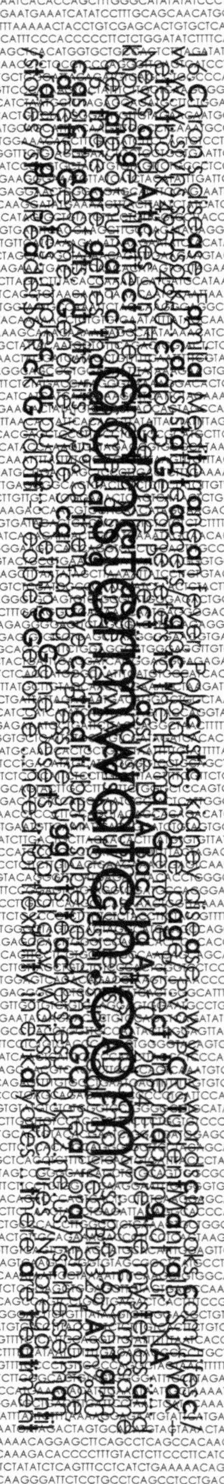

LET'S TAKE TIME TO COUNT

A

Eight million five hundred and sixty-two thousand four hundred and eighteen

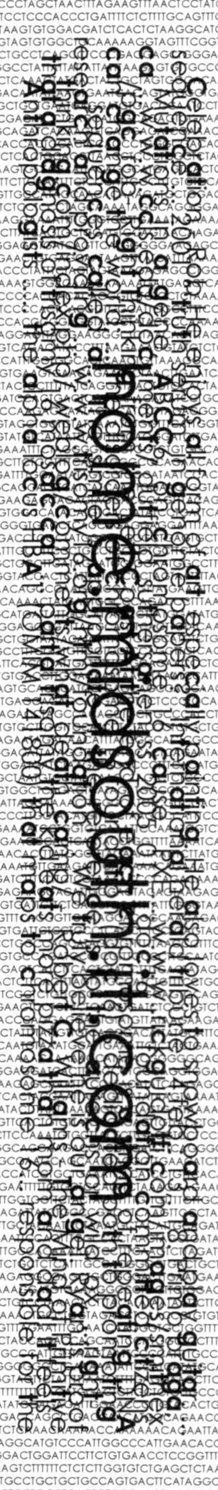

HOW IS THIS POSSIBLE?

H

viii

Eight million six hundred and fourteen thousand five hundred and thirty-five

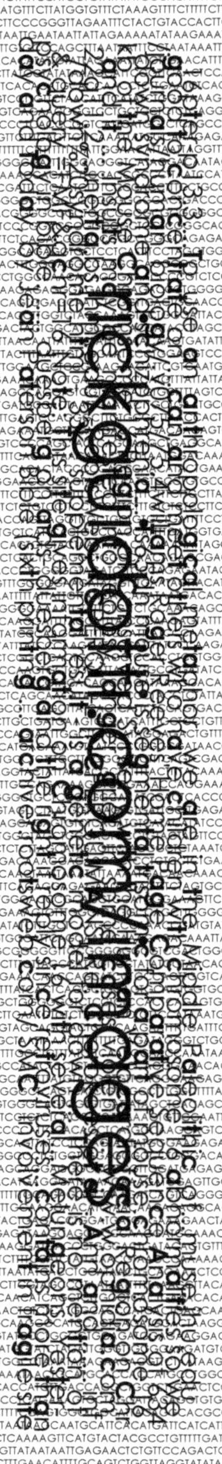

A NUMBER OF SYSTEMIC MANIFESTATIONS

R_x

Eight million six hundred and sixty-six thousand six hundred and thirty-three

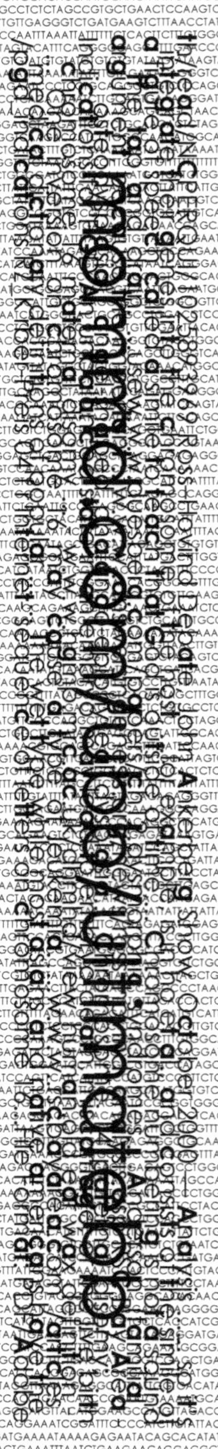

THE REAL ANSWER

M_{xi}

Eight million seven hundred and eighteen thousand four hundred and sixty-one

A CREATION OF THE DECEIVED

Eight million seven hundred and seventy thousand eight hundred and thirty-three

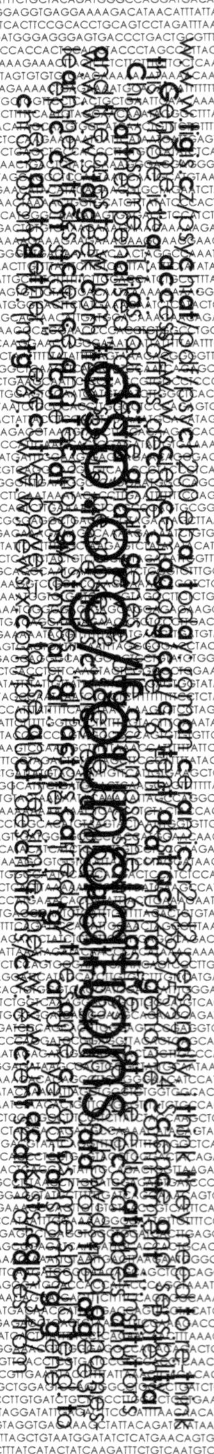

A FANCY HIGH GRAPHICS VERSION
E
x

Eight million eight hundred and twenty-two thousand two hundred and fifty-six

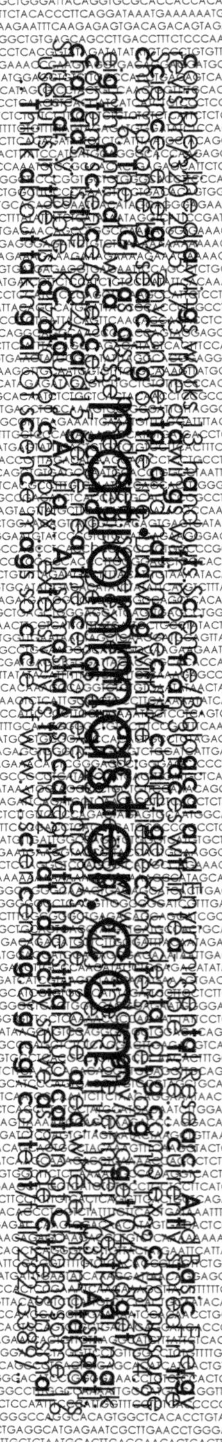

FROM A DEEP RED THROUGH TO BRIGHT COPPER

N

Eight million eight hundred and seventy-three thousand six hundred and forty

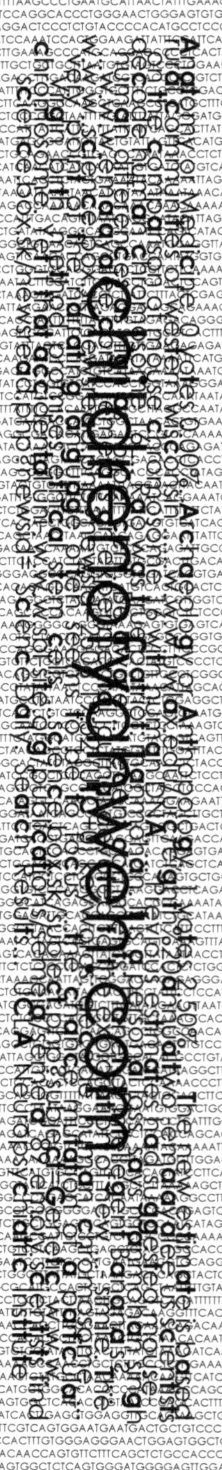

CORRECT BY THE LAW OF REASONING

C

xii

Eight million nine hundred and twenty-five thousand two hundred and ninety-one

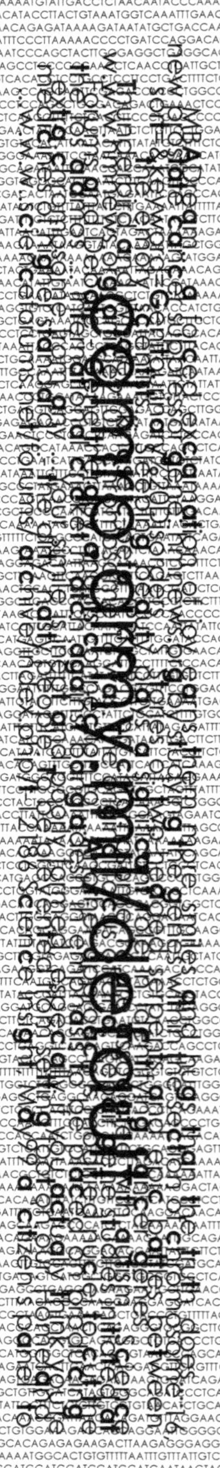

EVALUATION OF HYPERSPECTRAL IMAGING

C

vii

Eight million nine hundred and seventy-seven thousand one hundred and sixty-five

THE INTRINSIC DIFFICULTY OF TRYING TO DECONVOLUTE

W

Nine million twenty-nine thousand and sixty-one

CHOOSE AT LEAST TWO PRODUCTS TO COMPARE

B

xl

Nine million eighty-one thousand one hundred and thirty-one

FREE ESSAYS NEWER ESSAYS BETTER ESSAYS STRONGER

1

Nine million one hundred and thirty-three thousand four hundred and thirty-six

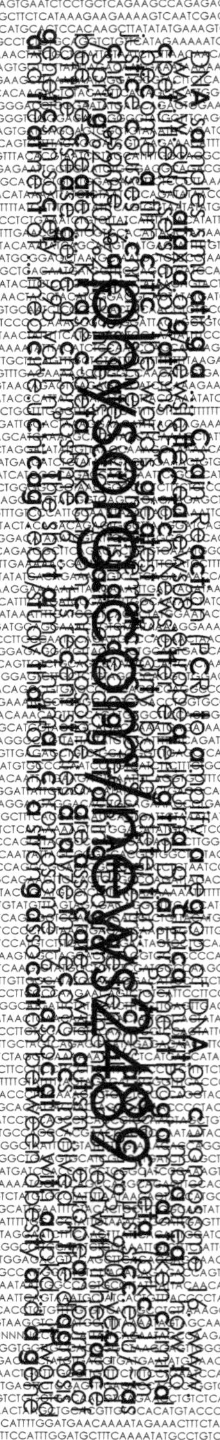

CONTROL THE INTRICATE FUNCTIONS

P

ix

Nine million
one
hundred
and eighty-
five
thousand
one
hundred
and ninety-
seven

A MAP OF AN IMMENSE COUNTRY

Nine million
two
hundred
and thirty-
seven
thousand
two
hundred
and six

NON PROFIT GENOMIC RESEARCH FUNDERS

G

CACCAAGTAGTGCTGTGATCTCAATAAATTGTGTCGGGTAGAAGTTCATCAAGTTGGGCAAAG [illegible]
CTTTATAAGTGTCTAACAAGGCCATGTTGCTCAAAGGAAGCAGAAATCTAAGCAGCAAGGGA [illegible]
TCCGAACGTTGTGTTAAATAAGAAATTTGGGATATAAGCTAAGAGGAGTGGGAGTCATAGAAAG [illegible]
CAGTGAAGGATAAAAAACGATGCTGGCTTTGACTAGGATGACGGAAGTAAAGACGGAAAGA [illegible]
GAGTGGATAGGTATATGGGTGATGGATGGGTGGATGGTTGAGTAGATGGATGGATGGGAAGAT [illegible]
TGAATGAATAGTTGGATGGATGGGTGAATGGGTGGATGGAAGAATCAGTGGAGGGGTAGATG [illegible]
ATAAATGGGTGGATAGACGAATGGGTAGATGGGTGAATGGGGGGGGGTAGATGAGTGAAT [illegible]
GAGTGAATGGATGAATGAATGAAAGGACTGGTGGGGGTGTATGTATGGATGGAAGGGTGAAT [illegible]
CAAGTAGATGGGTGGATACATGGATAAAAGGGTGGATTTGTGAGTGGATGTATGTGCTGATGG [illegible]
AATGGGTAGATGGGTAAGTGGTTGTATGAGTGGNNNNNNNNNNNNNNNNNNNNNNNNNN [illegible]
[illegible]

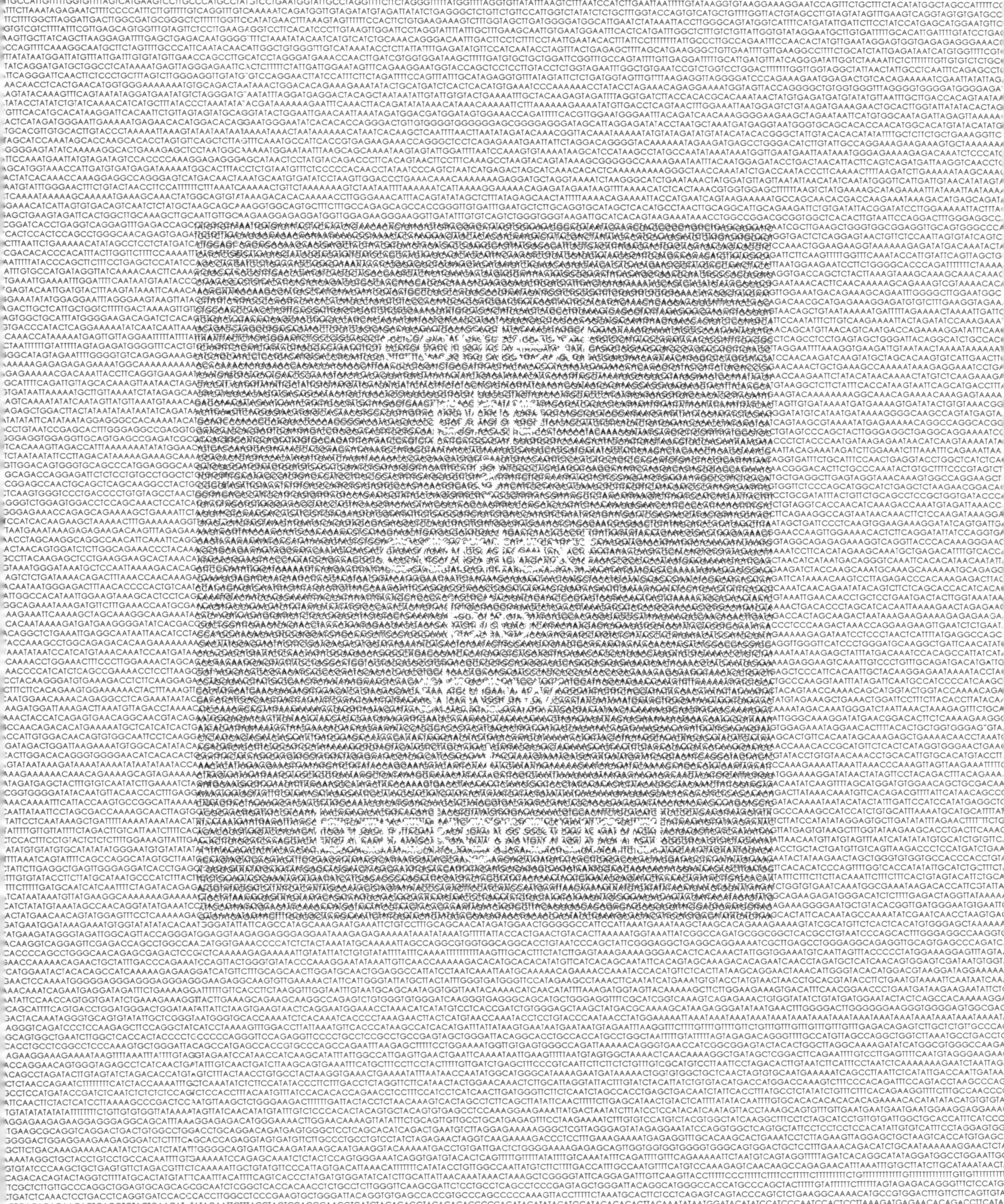

Nine million three hundred and sixty thousand eight hundred and fifty-eight

COMMUNICATE PRIVATELY WITH OTHER DIVERS

S IV

Nine million four hundred and twelve thousand eight hundred and forty-seven

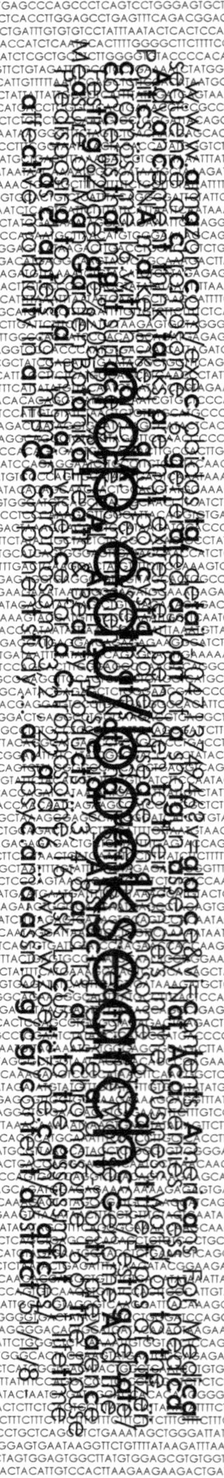

SORRY THERE IS A MISMATCH

N

Nine million
four
hundred
and sixty-
four
thousand
seven
hundred
and
seventy-
seven

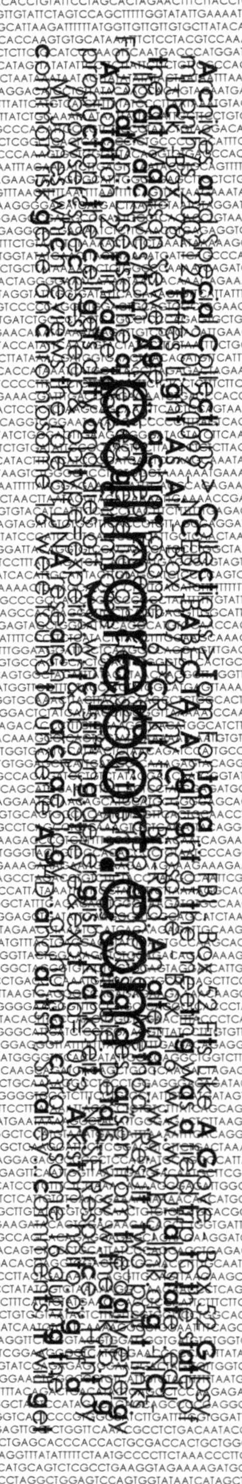

WHICH OF THE FOLLOWING STATEMENTS COMES CLOSEST

P

Nine million five hundred and seventeen thousand one hundred and sixty-seven

T

x1

GIVE AWAY ONE TRILLION ETEXT FILES

Nine million five hundred and sixty-nine thousand one hundred and forty-nine

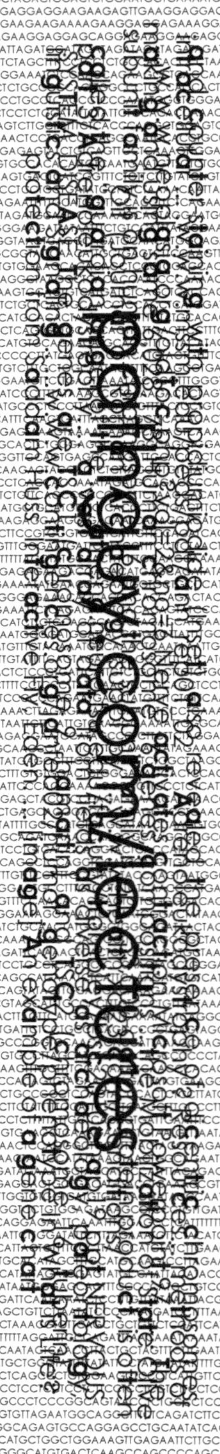

P
THE HELP YOU'RE LOOKING FOR IS PROBABLY HERE

Nine million six hundred and twenty thousand three hundred and twenty-six

G

WHAT'S MISSING MISASSIGNED AND MISASSEMBLED

Nine million
six hundred
and
seventy-two
thousand
four
hundred
and sixty-
three

J

Nine million seven hundred and twenty-four thousand two hundred and forty-six

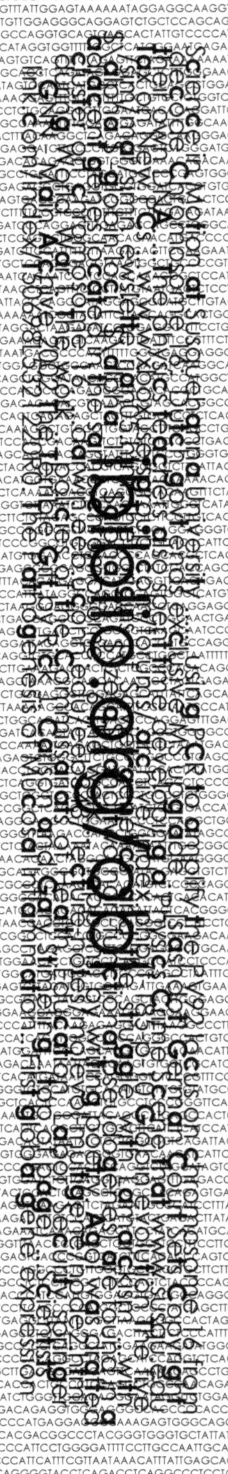

ENSURE PERPETUATION OF SOVEREIGNTY AT THE RISK TO LIVES

Nine million seven hundred and seventy-four thousand three hundred and sixty-two

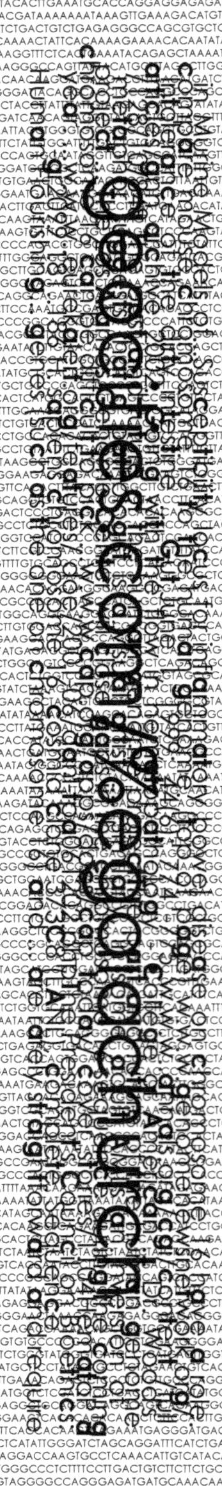

THIS IS A LARGE PREDATORY FAIRY SHRIMP

G

xiii

Nine million eight hundred and twenty-five thousand four hundred and twenty-one

THE INTEGRATION OF THE HEAD TO THE BODY

R viii

Nine million

eight

hundred

and

seventy-six

thousand

nine

hundred

and four

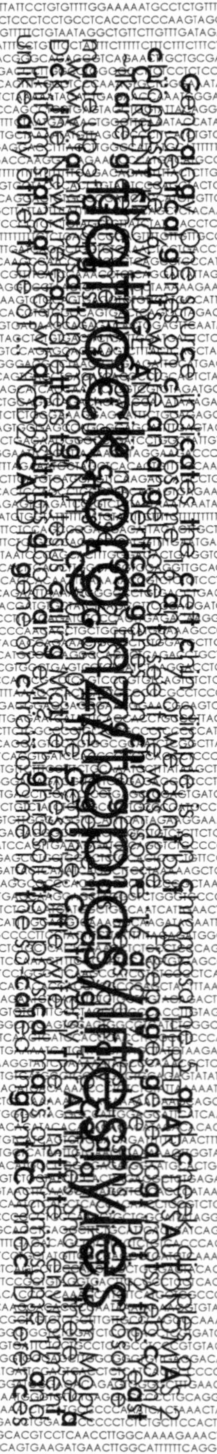

SOMETHING SMELLS BAD

Nine million nine hundred and twenty-seven thousand eight hundred and ninety-five

REVERSALS AND CONVERGENCES DO OCCUR

G

xvii

Nine million nine hundred seventy-eight thousand nine hundred and thirty-seven

P xi

COMPREHENDING METAPHORS WITH DIFFICULTY

Ten million twenty-nine thousand six hundred and forty-two

GROUND-BREAKING

Ten million seventy-nine thousand five hundred and twenty-two

G

HOUSE DUST MITES MOLDS COCKROACHES

Ten million one hundred and thirty thousand four hundred and fifty-seven

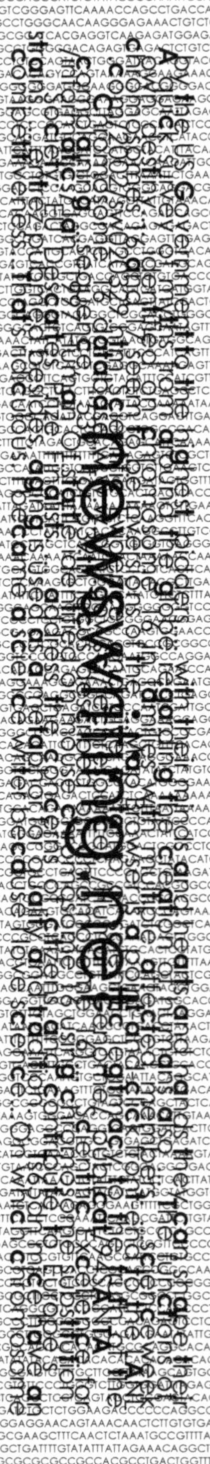

SOCIAL SMARTS STRATEGIES

N_x

Ten million one hundred and eighty-one thousand three hundred and nine

EROTICISING TRADEMARKED BATTLEMECHS

B

xiv

Ten million two hundred and thirty-two thousand and forty-nine

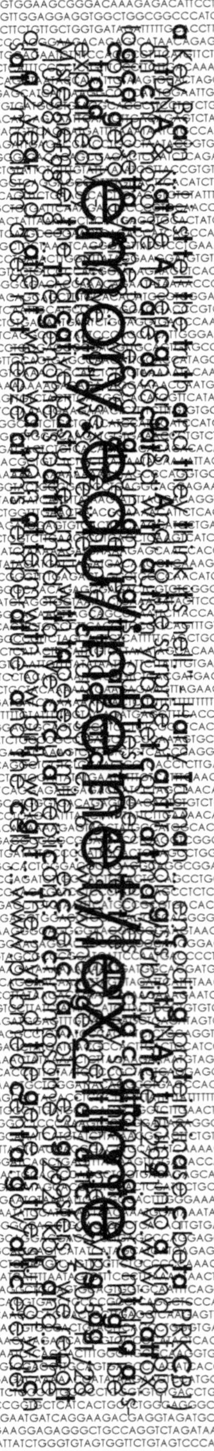

A MENTAL CODE OF AN HISTORICAL PERIOD

E iv

Ten million two hundred and eighty-two thousand three hundred and thirty-eight

THIS GENE IS YOUR GENE

D

Ten million three hundred and thirty-three thousand three hundred and eighteen

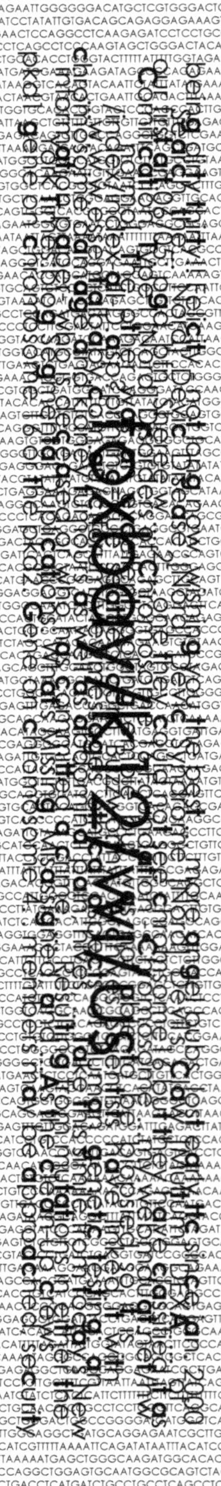

OBSERVE/ANALYSE COLOR VARIATIONS IN CORN SEED

F

vii

Ten million three hundred and thirty-three thousand three hundred and eighteen

DETECT DIM LINES

403

Ten million four hundred and thirty-five thousand four hundred and twenty-two

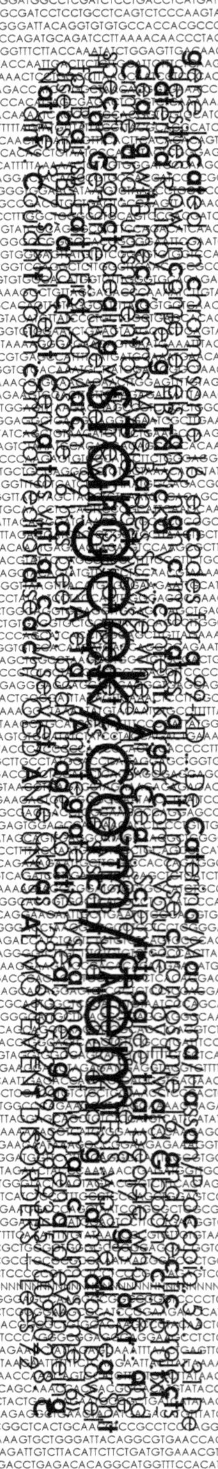

S

xii

CLICK HERE FOR INCREDIBLE AUTHENTIC CUBAN CIGARS

Ten million four hundred and eighty-seven thousand and sixty-seven

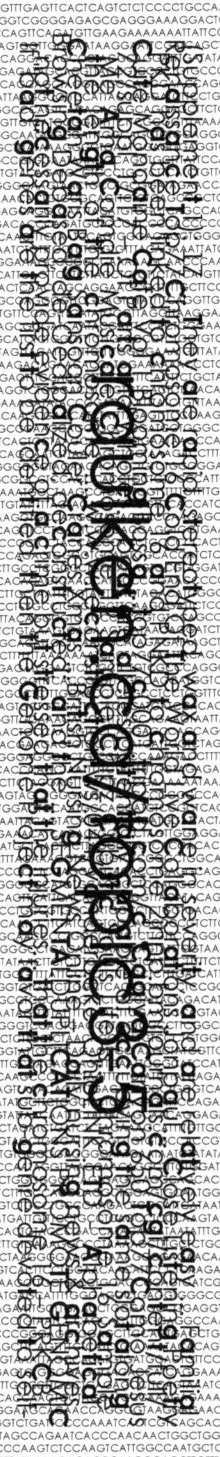

BRIGHT BIRTH OF A NEW BABY GALAXY

R

Ten million five hundred and thirty-eight thousand three hundred and fifty-two

PLUNGE INTO AN ALIEN WORLD

H

Ten million five hundred and eighty-nine thousand five hundred and twenty-four

A

MULTIPLE SOURCES THAT INFORM OR CONTEXTUALISE

Ten million six hundred and forty-one thousand two hundred and ten

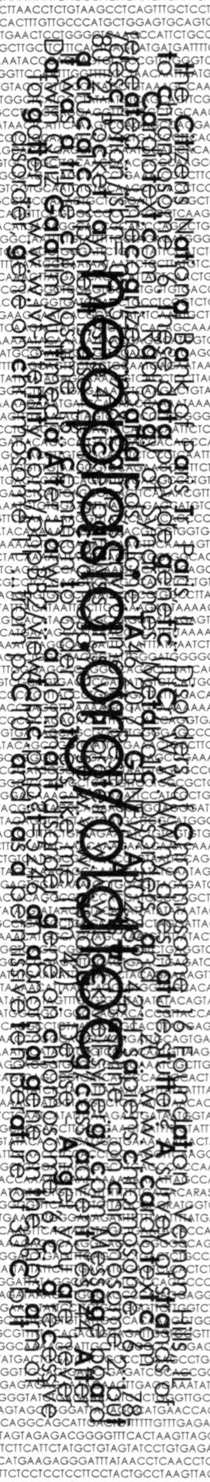

BREAKAGE/FUSION/BRIDGE CYCLES

N

vii

Ten million six hundred and ninety-three thousand two hundred and fifty-seven

THE GOOD TRANSLATOR CAN/SHOULD BE SENSITIVE

M

Ten million seven hundred and forty-five thousand eight hundred and eighty-nine

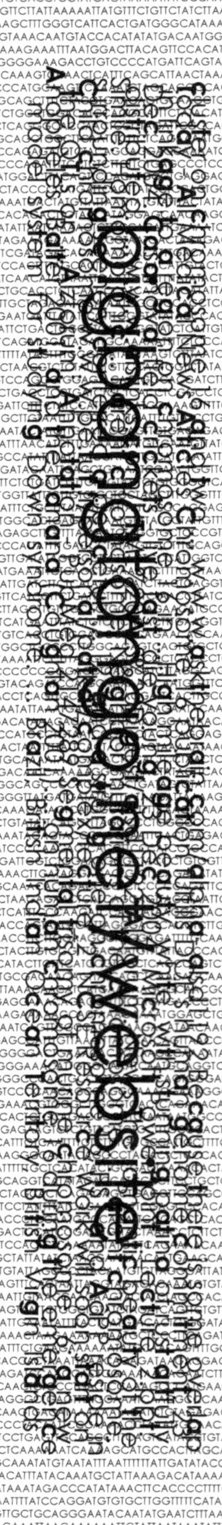

ONE GALAXY CLUSTER SWALLOWING ANOTHER

B

vi

Ten million seven hundred and ninety-seven thousand one hundred and ninety-seven

GROWTH PERFORMANCE CARCASS CHARACTERISTICS

C iv

Ten million eight hundred and forty-eight thousand six hundred and ninety

A NEW KIND OF CITIZENSHIP

L vii

Ten million nine hundred thousand eight hundred and twenty-five

STATE-OF-THE-ART CLINICAL TESTING

D

vii

Ten million nine hundred and fifty-three thousand one hundred and eighty-three

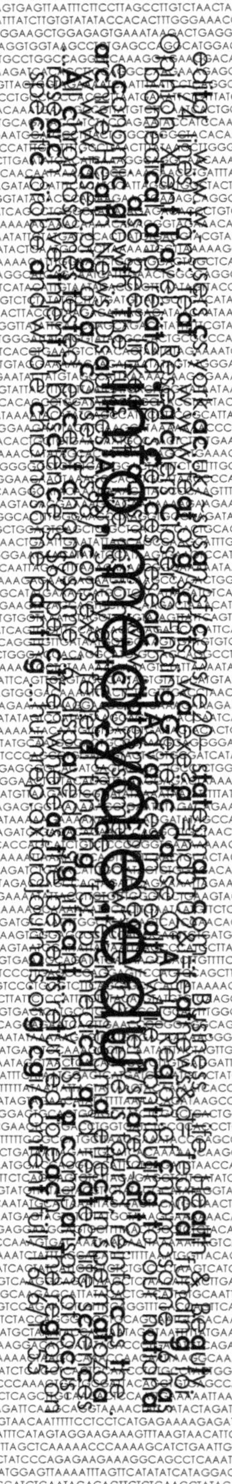

CLOSING IN ON THE MEN2A LOCUS

Eleven million five thousand four hundred and fourteen

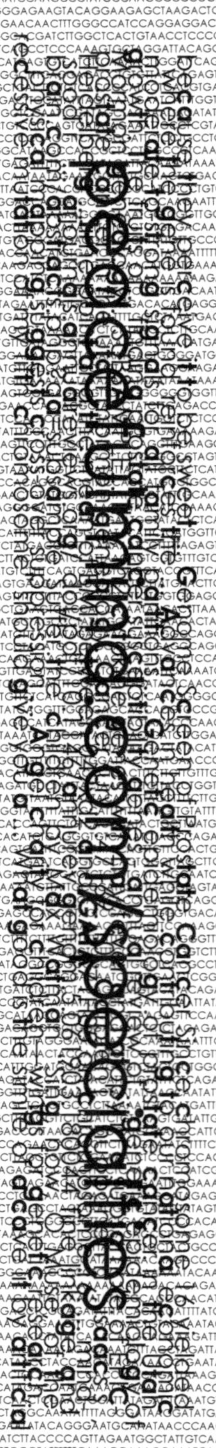

CRYSTAL GIFTS ARE PROFOUND
P

Eleven million fifty-seven thousand nine hundred and sixty-nine

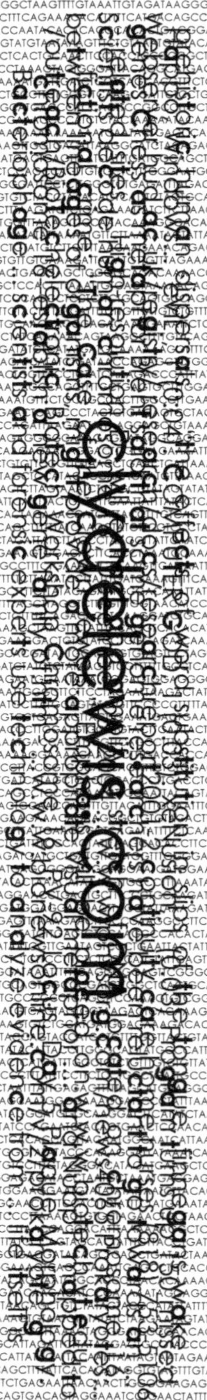

AND SHE HAS BEEN NAMED EVE

C

xiv

Eleven million one hundred and eleven thousand and sixty-two

MANY FALSE OVERLAPS HAVE BEEN DEDUCED

F

Eleven million one hundred and sixty-three thousand nine hundred and sixty-five

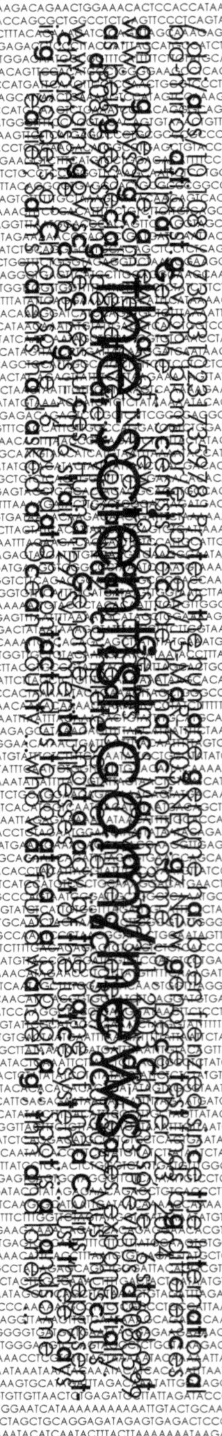

THE SCIENTIST MY SCIENTIST

Eleven million two hundred and sixteen thousand five hundred and twenty-two

STUDIES CAN "SHOW"

M

Eleven million two hundred and seventy thousand two hundred and fifty-three

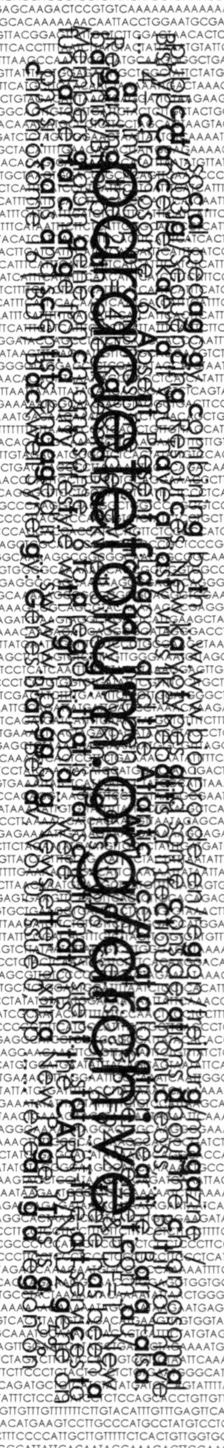

THE FIRST ORDER OF BUSINESS IS TO FIND

P

Eleven million three hundred and twenty-one thousand eight hundred and ninety-five

SUSTAINED REPETITIVE PATTERNED MOVEMENTS

D

viii

Eleven million three hundred and seventy-three thousand three hundred and three

A BRAIN IN A JAR

A

Eleven
million four
hundred
and twenty-
five
thousand
five
hundred
and fifty-five

IDENTIFICATION OF PRESUMPTIVE REGULATORY SITES

P

xiv

Eleven million four hundred and seventy-seven thousand two hundred and ninety-six

MORE THAN A BLUEPRINT

L

Eleven million five hundred and twenty-seven thousand nine hundred and five

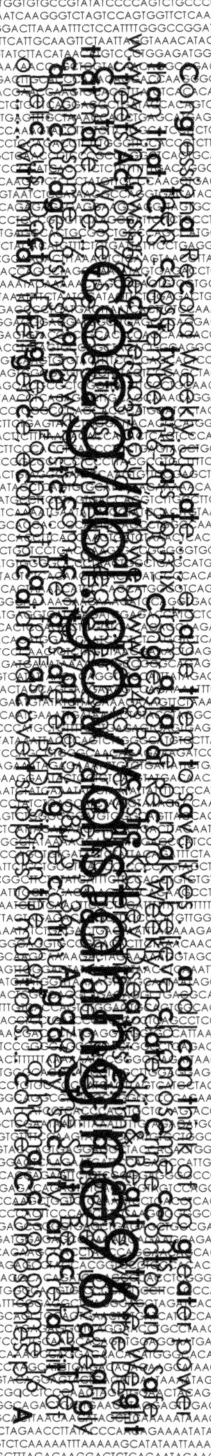

DESIGN DEVELOP IMPLEMENT AND DISTRIBUTE

C

Eleven million five hundred and seventy-nine thousand three hundred and fifty-one

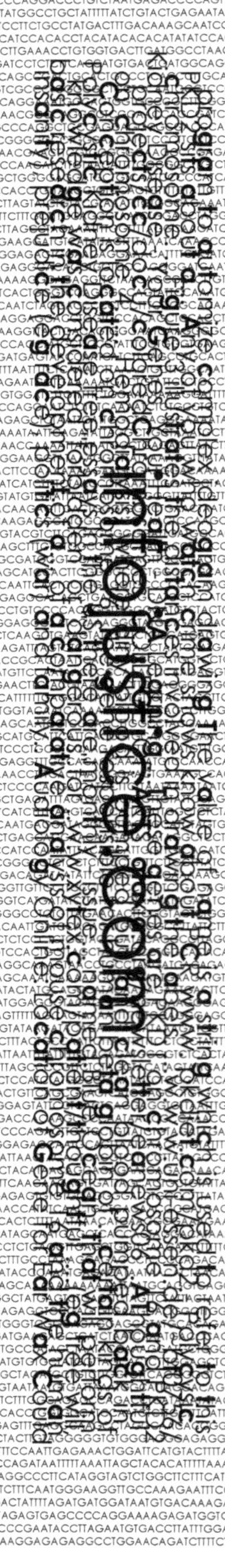

BE THE MOST INFORMED YOU CAN BE

Eleven million six hundred and thirty-one thousand and twenty-six

MEDIATE INTERACTIONS

Eleven
million six
hundred
and thirty-
one
thousand
and twenty-
six

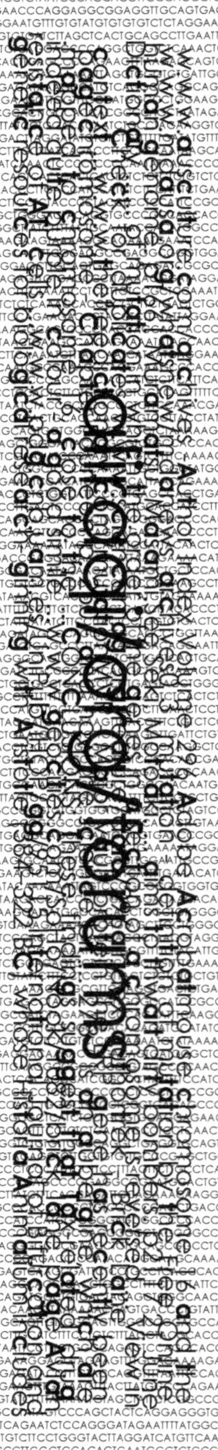

IF YOU DON'T LIKE THE OPINION DON'T REPLY

A

Eleven
million
seven
hundred
and thirty-
one
thousand
nine
hundred
and sixty-
four

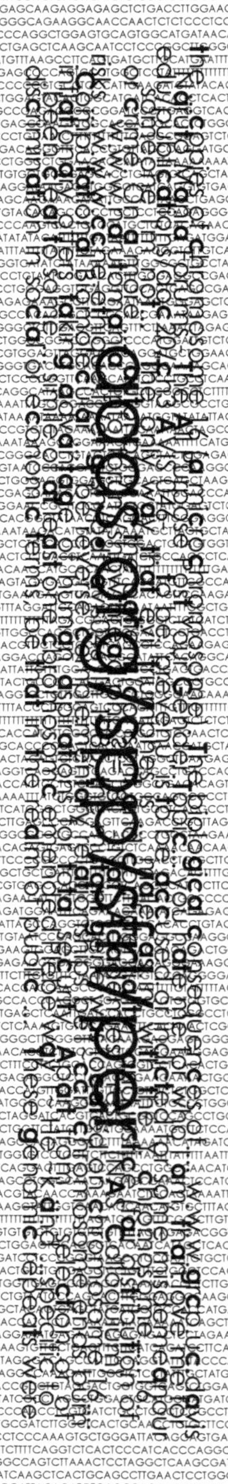

A

RELATED TO PROFESSIONAL ETHICS ISSUES

Eleven million seven hundred and eighty-five thousand two hundred and thirty-three

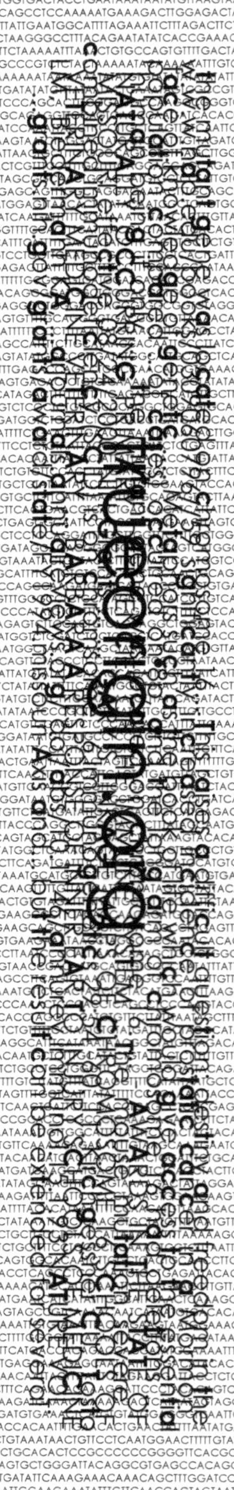

T

xiv

SELECTIVE USE OF DATA IS NO SUBSTITUTE FOR EVIDENCE

Eleven million eight hundred and thirty-eight thousand two hundred and fifteen

WHAT'S YOUR SPIRITUAL TYPE ?

B

Eleven million eight hundred and sixty-eight thousand six hundred and forty-three

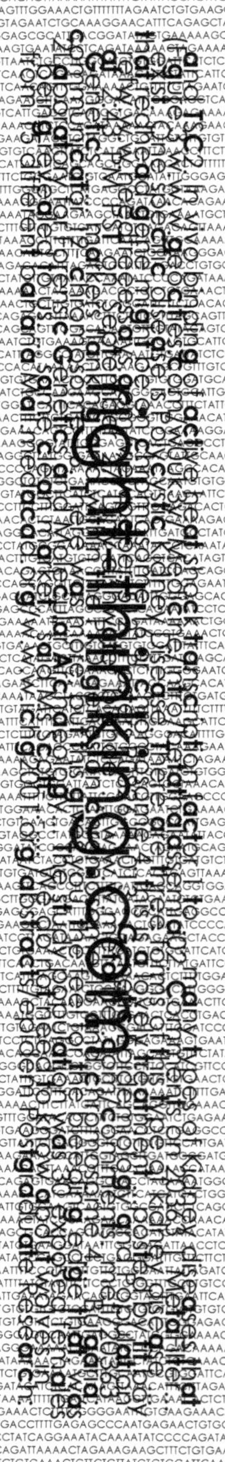

WHAT DEATH CAMP?

R x

Eleven million nine hundred and nineteen thousand eight hundred and fifty-two

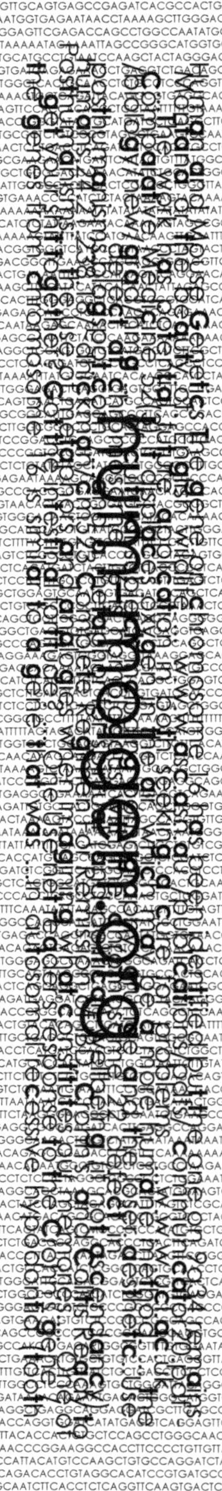

LEARN ABOUT THE BENEFITS OF BANNER ADVERTISING

H

Eleven
million nine
hundred
and seventy
thousand
three
hundred
and forty-
seven

BEFORE YOU START TO WRITE THINK DEEPLY

C xx

Twelve
million
twenty-two
thousand
four
hundred
and nine

THINK ABOUT DISTRIBUTED COGNITION IN HISTORICAL TERMS

P xii

Twelve million seventy-two thousand nine hundred and ninety-seven

A GENERIC MOTIF

B

Twelve million one hundred and twenty-four thousand four hundred and forty-two

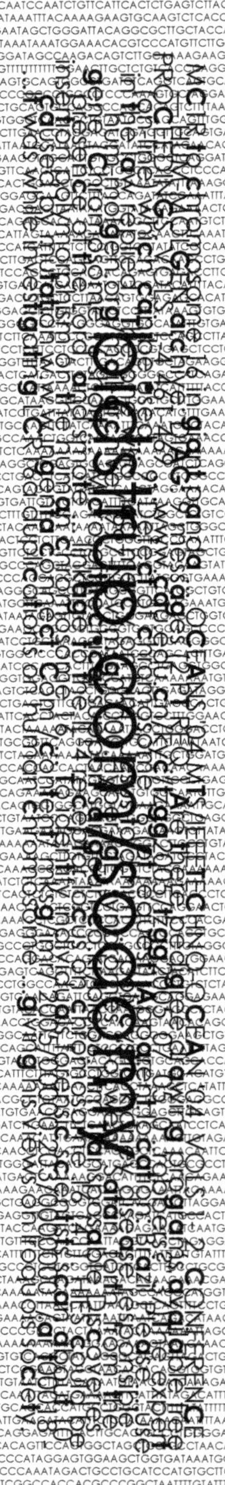

BISON GAZELLES ANTELOPE SAGE GROUSE

B

Twelve million one hundred and seventy-four thousand nine hundred and ninety-six

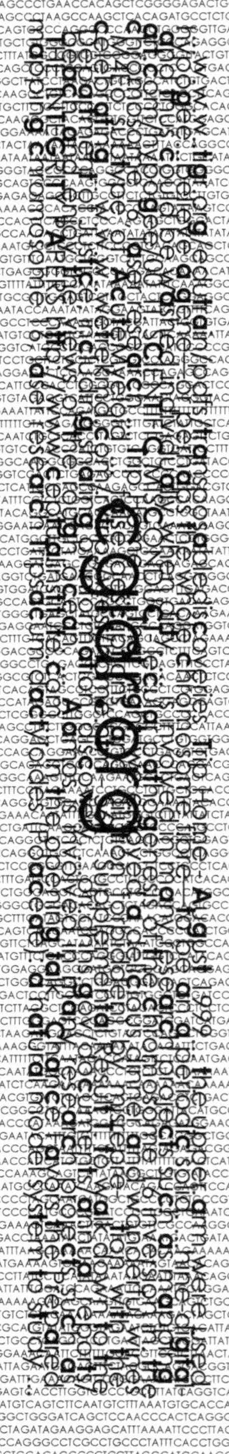

C

DESERTIFICATION DROUGHT POVERTY AND AGRICULTURE

Twelve million two hundred and twenty-six thousand seven hundred and eleven

A LIBRARY IS A RANDOM ASSORTMENT OF CLONES
U

Twelve million two hundred and seventy-seven thousand one hundred and eighty-five

AN AGENDA IS MADE

N

Twelve million three hundred and twenty-eight thousand four hundred and thirty-four

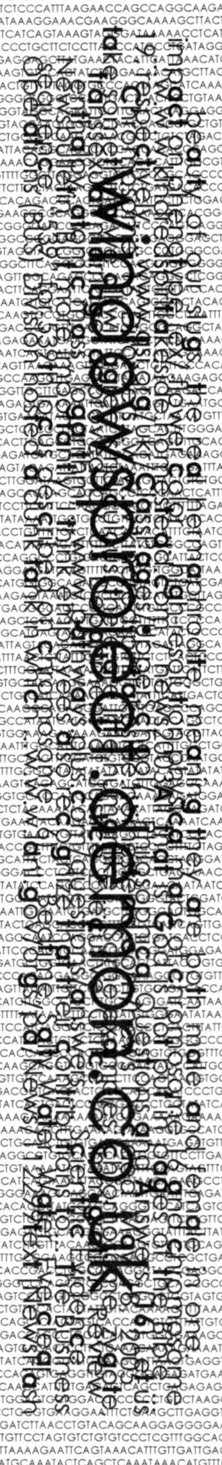

GAMES SEEM IRRESISTIBLY TO LEAD TO "NONSENSE"

W

Twelve million three hundred and seventy-eight thousand nine hundred and forty-four

SYMBOL ALIASES CYTO LOCATION

G

Twelve
million four
hundred
and fifty
thousand
three
hundred
and thirteen

IS FUGU NOT A POISONOUS FISH?

F

xi

Twelve million four hundred and eighty-three thousand two hundred and eight

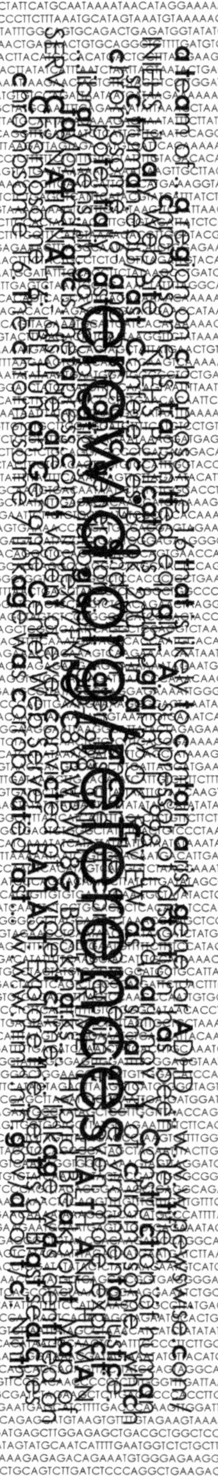

CLARIFY WHAT IS NOW KNOWN

E

Twelve million five hundred and thirty-six thousand two hundred and seventy-five

sff.net/people/jcavelos/bad

SOME ALIEN QUALITY

S

viii

Twelve million five hundred and eighty-nine thousand nine hundred and ninety-six

THE GOAL IS TO WRITE US SOME POEMS

A

vii

Twelve million six hundred and forty-two two thousand two hundred and eighty-six

NEWS FROM THE COMMUNITY

K

Twelve million six hundred and forty-two thousand two hundred and eighty-six

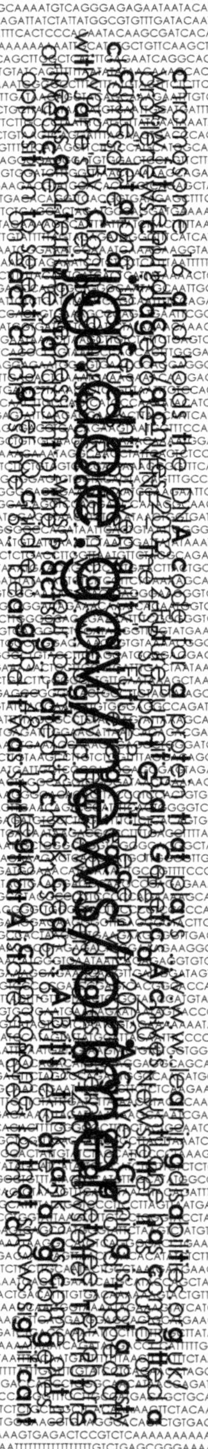

HYPOTHESIS GENERATION AS OPPOSED TO ROTE EDITING

J

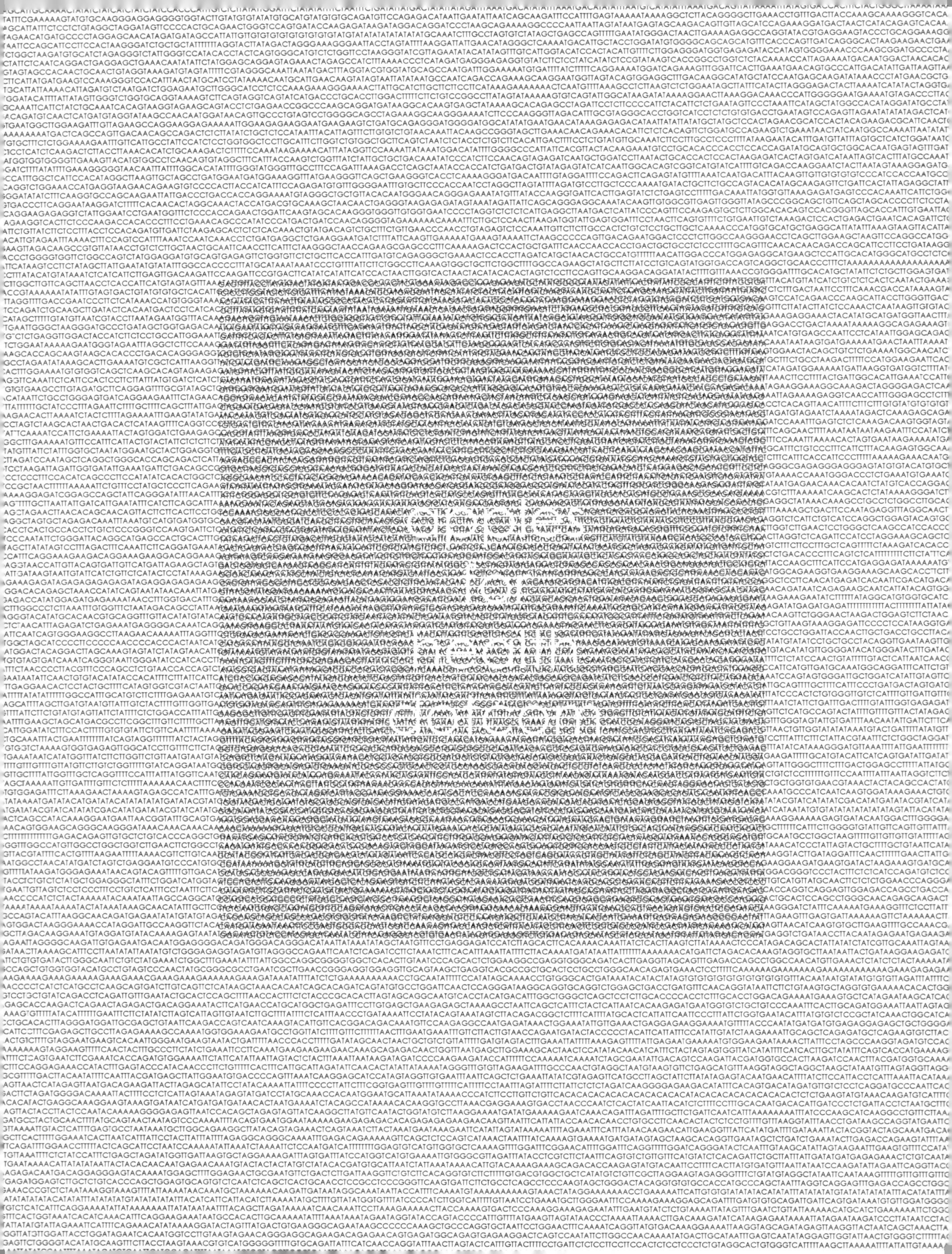

[illegible]

[illegible]
TTAAGTTGGAAAATTCATATTACACCACCACAAGACTTATAGAGTAAGACACTGTGGTGTTGGTGAAAGAAGATAGGGATTAAAAGAACAGACCATAAGGAATTGCACTGTATTTCAAGTTTGATGTAGCTGTAGTTGTCAGAGGATTCAAATTTTCATAAGTTTTTCTTTTCTTCGTTTATTTTTCTCTT
TTTGCTTCCTTCCATTCTTTTTTGTTTTCCCTGTTGCCTTTGGAGTTTCCTAAGTCCTCCTCCTCCTTAGAGATAGTCTGTAGCTTGAAGCTTTTTCATCTATAATCCACCATAACCATATTGGAGTCCTATTAGTGTAGAAGTAAGATGTGGGGGTGGAAGAGTATTTCACAAACTTCTGATTAACTCTCAAG
AACTGTGTCTCCAGTGTGACCTTCACAATTTTCTCTTTAGAGATGTGGCTTCTTTTTTTGCACCCACTATCTACCCCTTTCTTTGACTGTAGCATTCTCAATCAATTATCTTTGAAACTAACAACCCTGTTGACTGTGACTTTGCCCCTTTCCTCGGTAAGACAGCAAGGCTGGAAGGGGTTGGAGTGGG
CCTAGGCTGGATGATTTTATGAAATAGAACTTTCTTCTGGAGAGTACGCATTTGTTAAGGAGAAGGTTATTAGTGTATTTTCACCGTGATTACTCTTCCTCTTTTGTTTTTAGAGCCAGGAGGGGATATTTGTGAATTATCCACTGGTGAAGTTTCTGGTGATAGAGCGCAGGAAGGTATGGAGGCCCT
CCCACCCATATAGCCCTGGGAGACTCTCACTTTCAGACTAGTTCACCCTTCACTCTAGCAATTCATTGACATTAGGAAGGATTTGTTGCTAGCAGTTTGTGGACCTAGCAGCTTCTACCCCTGGTAAGTGGATCACAATATCTCTCTAGCTGTGTTTGTTTCTCCAGAGTACACTGTGGCAGTTTT
CTTACTTCTCTGGTGTGTCTGAGAAAAGTCGATTTTCAGCTTGTGTAATTTTTTGTTGTTGCTGTAAAGGTGGGCAGCTCTTTACATATTGAAGCTGTAACCTGAAATCCATTATTTTTTTATTATACAATGTTAGCTCTAGGTGTAATGTAAAAGTAACAGATGCATGGTCTATTTTATTATCTTCTAACTCTCCA
CATATCACTTTATGGAGTTCAAACTACTTTAAAGAGTTTATGCAATTAAATTCTTTCTTGTTGTATGAACATAAGCTTTTGATCATTTTCATAGTAATCTAAACATTTGAAACATAAAATACAAAATTGAACTTATTCACCTTACCTACTTTAAATGGATATATAATTATGTCTTTTTTCTGAAATAATATCTTTAGTCA
TTTTTTCAGCTTTCTGTTTTTTAATTTTGATATTATTTTCCCCACTTGAATTCCTAAGTATAGAAATAGAATCCCATTATTAAAATGCTGATATCTGACTAGATTATCTCTAAAGAGTATTACAAGACCCAGTATTTCATGATTCTATAGTCGTTCTATATTTCTTTTATACTCAGAGTGTGAACTGGATCTTAGTTATA
ATCTAGTTTATTTCAGGATGTTGGTAGAAACGGGGACAAATTATTCACCAATCTTCATCCCAGGCTCAGAGGAAGATTATACATCCAAGTTTTCCACTATAATTAGGTTGGAGCCATATGACTAGGCTCTGGACAAAGGACAGTGGAAATACTAGGCATCACTTTTAGGCATGCCACACCTT
GCCTCCAAGACCCCTTAAAAATGTTAATGATTATACTCTCACAGCAGGAAGCAACATATGGTCAATGGGTGTGTGTGTGTGTTTGTGTGTGT [illegible]
[illegible]

Twelve million eight hundred and twenty-one thousand seven hundred and eight

LOOK AT THE ELEGANCE
A
xvii

Twelve
million eight
hundred
and
seventy-five
thousand
three
hundred
and twenty-
two

WOW THAT'S JUST BULLSHIT

M viii

Twelve million nine hundred and twenty-eight thousand six hundred and thirty

LEARN HOW A PATHOLOGIST MAKES A DIAGNOSIS

Twelve million nine hundred and eighty-two thousand five hundred and eighty-six

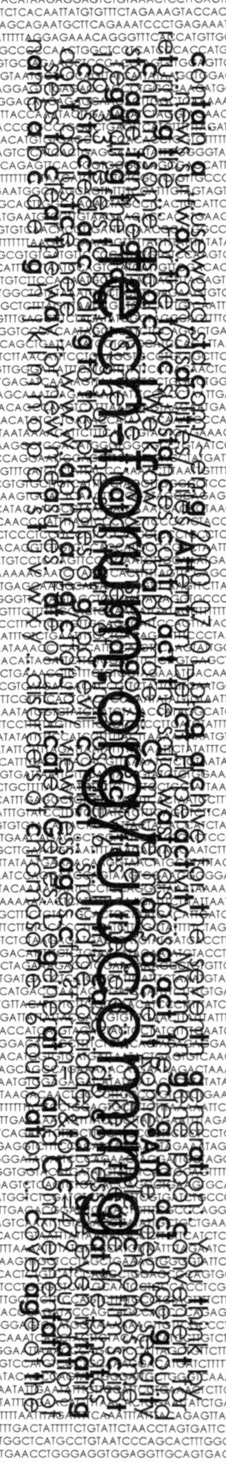

T

WE CARE VERY MUCH ABOUT WHAT YOU HALF TO SAY

Thirteen million and thirty-five thousand five hundred and seventy-one

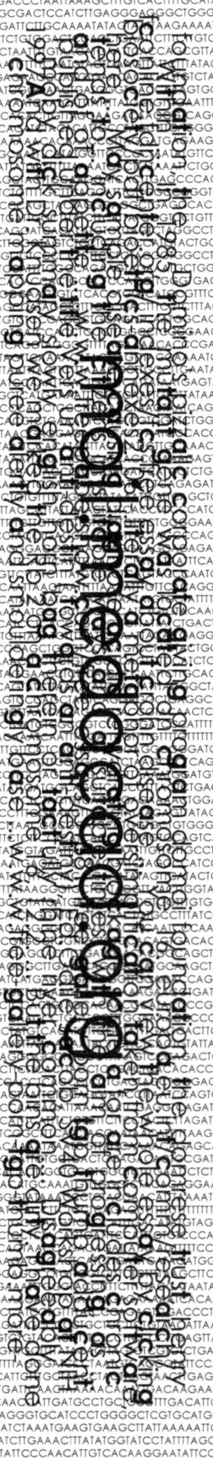

THE RESULTS MAY BE MISUNDERSTOOD

M

Thirteen million and eighty-eight thousand seven hundred and ninety-six

ANSWER "SOME OF THE MOST FREQUENTLY ASKED QUESTIONS"

S

fx

Thirteen million one hundred and forty-one thousand seven hundred and fifty-two

A FREEWILL OFFERING CATERED MEAL WILL FOLLOW

Thirteen million one hundred and ninety-five thousand and eighty-six

SEPARATE SPACES OF DESCRIPTION

C xiii

Thirteen million two hundred and forty thousand and thirty-three

YOU CAN SEE THE ELECTROPHRETIC PATTERN

A

xiii

Thirteen million three hundred and one thousand six hundred and twenty-one

N

xiii

HIGHLY MODULARIZED AND ABSTRACTED

Thirteen million three hundred and fifty-three thousand four hundred and eight

A VARIETY OF ODDITIES AND OTHER ENDS

M

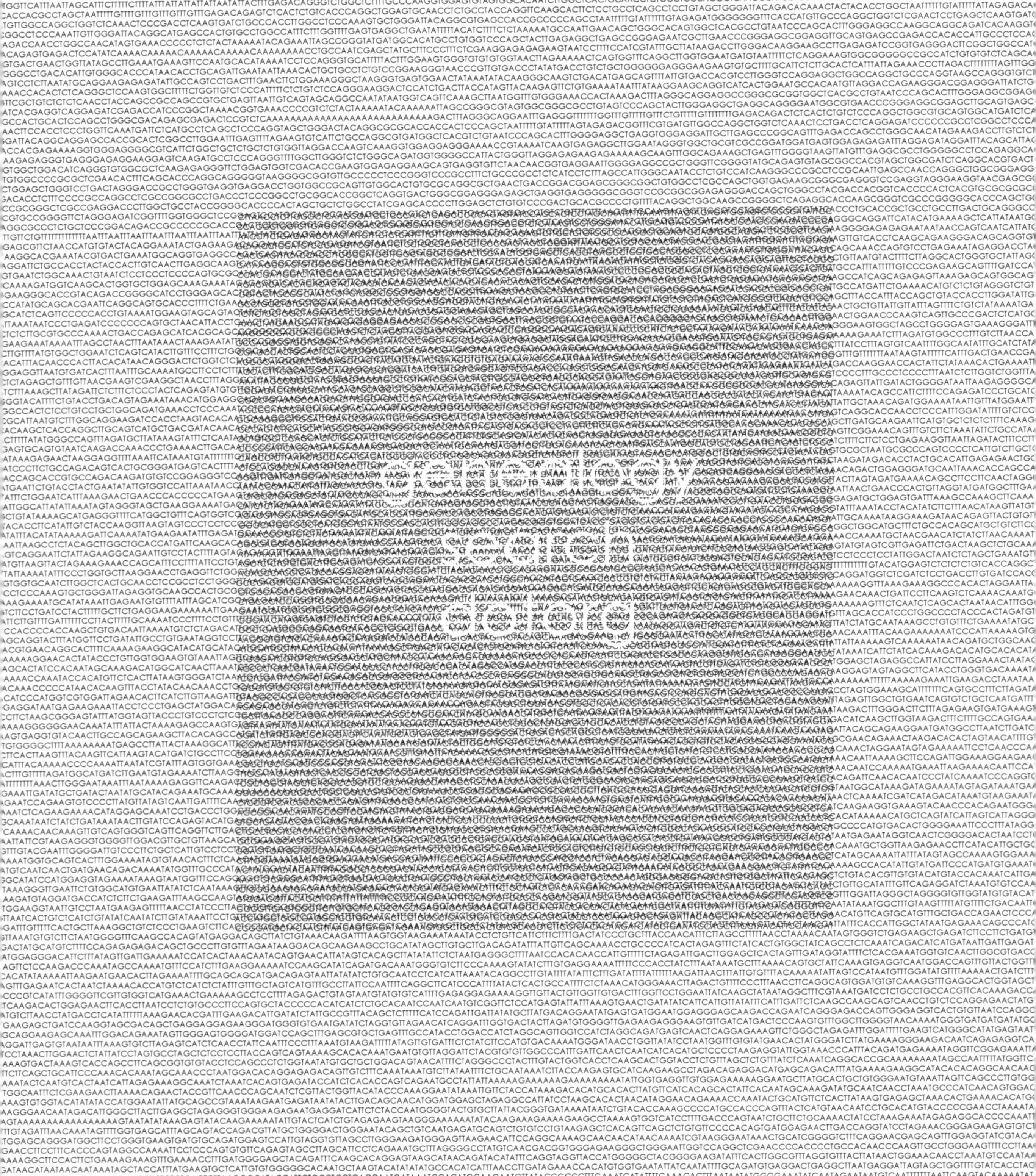

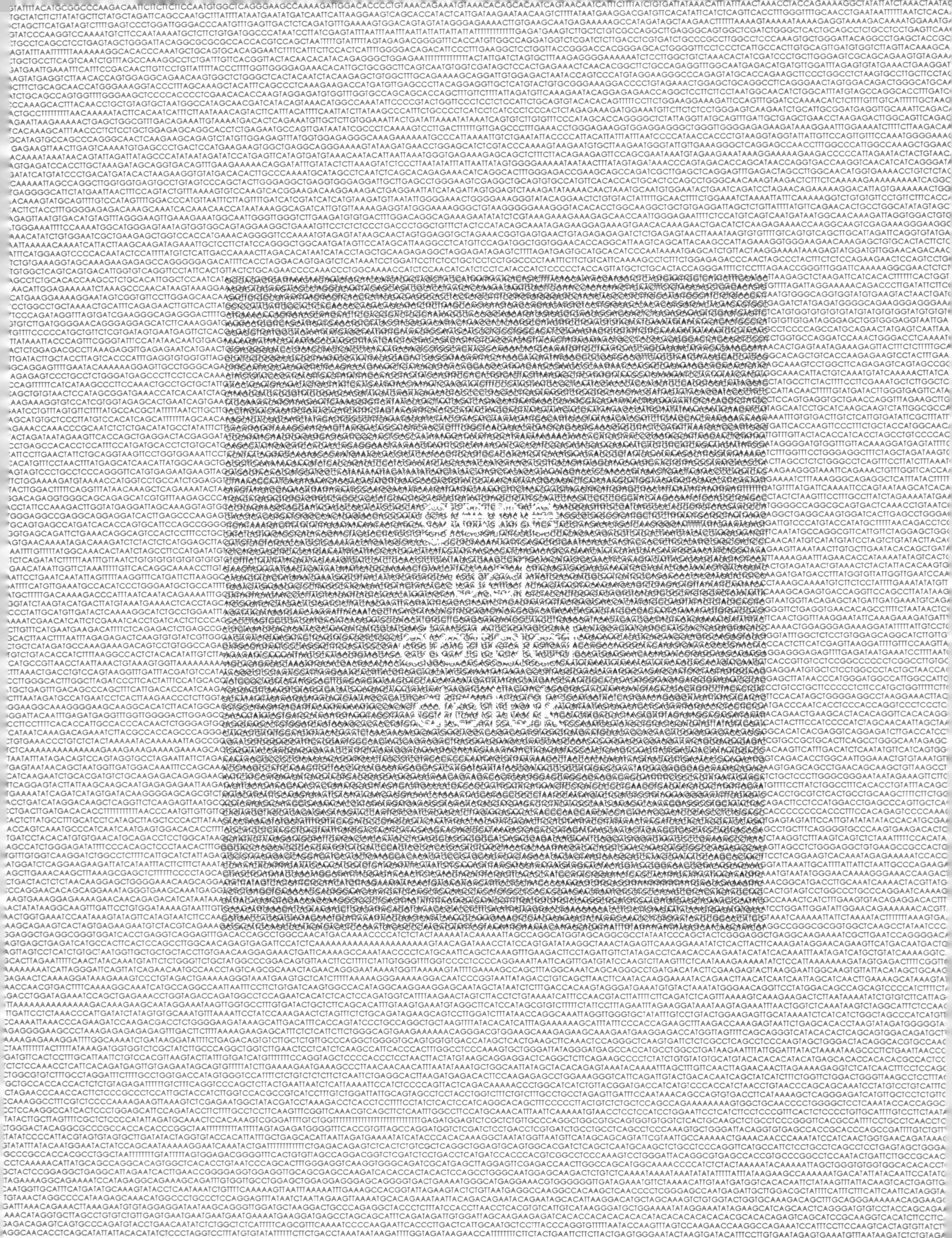

CACATGCACCTTCCCATAAATCCTGACCCTGCATCAGGGCCTGGTCTCTGGCAGCCTCCTGGGATGGGTGCCACCCCAGCCAGAGCCCGAAGGCGGAGCTCACCTGACACTGCCCAGGGGGCTGCGCTTCTCCTGCTCCTGCCGCCGGGCTCGCAGCTGCTCCTCCTCGC
CTCCTCCTCCATGGCAGAGGCTCCTCTTGGCCCGACGACGGTTCTCCTATTGAAGAGGGACAGACCTGGCTGAGCCTACTGGGGACCACTAAGGAAGGAGGCTCGTCGGCACCCAAGTCCTGCCACAAACCAGGTGCTGCATCCACCAGGCCAGTGCTCCACATGGCT

[illegible]

Thirteen
million five
hundred
and eighty-
eight
thousand six
hundred
and eighty-
four

THE FOLLOWING REFERENCES HAVE BEEN REMOVED FROM THE BODY OF EBOOK #2216

ref|NT 000673|Hs16 175| Homo sapiens 16p13.3 sequence /len=81486*G*>ref|NT 000657|Hs16 162| Homo sapiens 16p13.3 sequence /len=166010*
ref|NT 000658|Hs16 163| Homo sapiens 16p13.3 sequence /len=64089*>ref|NT 000659|Hs16 1511| Homo sapiens 16p13.3 sequence
/len=79586*>ref|NT 000660|Hs16 1512| Homo sapiens 16p13.3 sequence /len=78941*>ref|NT 000661|Hs16 165| Homo sapiens 16p13.3 sequence
/len=302711*>ref|NT 000662|Hs16 166| Homo sapiens 16p13.3 sequence /len=66544*>ref|NT 000663|Hs16 167| Homo sapiens 16p13.3 sequence
/len=121193*>ref|NT 000664|Hs16 168| Homo sapiens 16p13.3 sequence /len=124604*>ref|NT 000667|Hs16 170| Homo sapiens 16p13.3 sequence
/len=101349*>ref|NT 000668|Hs16 1514| Homo sapiens 16p13.3 sequence /len=35886*>ref|NT 000669|Hs16 171| Homo sapiens 16p13.3 sequence
/len=119642*>ref|NT 000670|Hs16 172| Homo sapiens 16p13.3 sequence /len=145784* ref|NT 000671|Hs16 173| Homo sapiens 16p13.3 sequence
/len=215441*>ref|NT 000672|Hs16 174| Homo sapiens 16p13.3 sequence /len=82935*>ref|NT 000676|Hs16 178| Homo sapiens 16p13.3 sequence
/len=116849*>ref|NT 000677|Hs16 179| Homo sapiens 16p13.3 sequence /len=162822*>ref|NT 000678|Hs16 180| Homo sapiens 16p13.3 sequence
/len=47644*>ref|NT 000679|Hs16 181| Homo sapiens 16p13.3 sequence /len=264272*>ref|NT 000680|Hs16 182| Homo sapiens 16p13.3 sequence
/len=169765*>ref|NT 000681|Hs16 183| Homo sapiens 16p13.3 sequence /len=164538>ref|NT 000706|Hs16 208| Homo sapiens 16p12 sequence
/len=138839**>ref|NT 000682|Hs16 184| Homo sapiens 16p13.13 sequence /len=35641*>ref|NT 000683|Hs16 185| Homo sapiens 16p13 sequence
/len=53060*>ref|NT 000684|Hs16 186| Homo sapiens 16p13.1 sequence /len=139376*>ref|NT 000685|Hs16 187| Homo sapiens 16p13.3 sequence
/len=80831*>ref|NT 000686|Hs16 188| Homo sapiens 16p13 sequence /len=215286*>ref|NT 000687|Hs16 189| Homo sapiens 16p13 sequence
/len=236177*>ref|NT 000688|Hs16 190| Homo sapiens 16p13 sequence /len=512439*>ref|NT 000689|Hs16 191| Homo sapiens 16p13 sequence
/len=167880*>ref|NT 000690|Hs16 192| Homo sapiens 16p13.11 sequence /len=145831*>ref|NT 000691|Hs16 193| Homo sapiens 16p13.11 sequence
/len=127485*>ref|NT 000692|Hs16 194| Homo sapiens 16p13 sequence /len=153773* ref|NT 000693|Hs16 195| Homo sapiens 16p13 sequence
/len=237215*>ref|NT 000694|Hs16 196| Homo sapiens 16p12 sequence /len=87077*>ref|NT 000695|Hs16 197| Homo sapiens 16p12 sequence
/len=218074*>ref|NT 000696|Hs16 198| Homo sapiens 16p12 sequence /len=77713*>ref|NT 000697|Hs16 199| Homo sapiens 16p12.3 sequence
/len=86156*>ref|NT 000698|Hs16 200| Homo sapiens 16p11-p13 sequence /len=511223*>ref|NT 000699|Hs16 201| Homo sapiens 16p12 sequence
/len=106760*>ref|NT 000700|Hs16 202| Homo sapiens 16p13.1-p12 sequence /len=265571*>ref|NT 000701|Hs16 203| Homo sapiens 16p12 sequence
/len=307828*>ref|NT 000702|Hs16 204| Homo sapiens 16p12 sequence /len=173882*>ref|NT 000703|Hs16 205| Homo sapiens 16p12 sequence
/len=284808*>ref|NT 000704|Hs16 206| Homo sapiens 16p12 sequence /len=128765*>ref|NT 000705|Hs16 207| Homo sapiens 16p12 sequence /len=134743*
ref|NT 000707|Hs16 209| Homo sapiens 16p13.1 sequence /len=93431*>ref|NT 000708|Hs16 210| Homo sapiens 16p12.2 sequence
/len=259894*>ref|NT 000709|Hs16 211| Homo sapiens 16p12-p13.1 sequence /len=361235*>ref|NT 000710|Hs16 212| Homo sapiens 16p12 sequence
/len=103911*>ref|NT 000711|Hs16 213| Homo sapiens 16p12.2-p12 sequence /len=213633*>ref|NT 000712|Hs16 214| Homo sapiens 16p11.2-p12 sequence
/len=163216*>ref|NT 000713|Hs16 215| Homo sapiens 16p11-p12 sequence /len=139480*>ref|NT 000714|Hs16 216| Homo sapiens 16p11-p12 sequence
/len=134450*>ref|NT 000715|Hs16 217| Homo sapiens 16p12 sequence /len=131530*>ref|NT 000716|Hs16 218| Homo sapiens 16p12 sequence
/len=114411*>ref|NT 000717|Hs16 219| Homo sapiens 16p11 sequence /len=174582*>ref|NT 000718|Hs16 220| Homo sapiens 16p11.2 sequence
/len=227123*>ref|NT 000719|Hs16 221| Homo sapiens 16p11.2 sequence /len=105485*>ref|NT 000720|Hs16 222| Homo sapiens 16p12-p13 sequence
/len=240805*>ref|NT 000721|Hs16 223| Homo sapiens 16p11 sequence /len=139958*>ref|NT 000722|Hs16 224| Homo sapiens 16p11.2 sequence
/len=319010*>ref|NT 000723|Hs16 225| Homo sapiens 16p11.1-p11.2 sequence /len=234542*>ref|NT 000724|Hs16 226| Homo sapiens 16p13 sequence
/len=70048*>ref|NT 000725|Hs16 227| Homo sapiens 16p12 sequence /len=130020*>ref|NT 000726|Hs16 228| Homo sapiens 16p13-q23 sequence
/len=82883*>ref|NT 000727|Hs16 229| Homo sapiens 16p11.2 sequence /len=120955*>ref|NT 000728|Hs16 230| Homo sapiens 16p11.1-p11.2 sequence
/len=161580*>ref|NT 000729|Hs16 231| Homo sapiens 16p11.2 sequence /len=150296*>ref|NT 000730|Hs16 232| Homo sapiens 16p11.2-p12 sequence
/len=69015*>ref|NT 000731|Hs16 233| Homo sapiens 16q23 sequence /len=112235*>ref|NT 000732|Hs16 234| Homo sapiens 16q13-q21 sequence
/len=217873*>ref|NT 000733|Hs16 235| Homo sapiens 16q21 sequence /len=110209*>ref|NT 000734|Hs16 236| Homo sapiens 16q21-q22 sequence
/len=101311**>ref|NT 000735|Hs16 237| Homo sapiens 16q21-q22 sequence /len=118521*>ref|NT 000736|Hs16 238| Homo sapiens 16q21 sequence
/len=130491*>ref|NT 000737|Hs16 239| Homo sapiens 16q22-q23 sequence /len=259474*>ref|NT 000738|Hs16 240| Homo sapiens 16q22.2 sequence
/len=117435*>ref|NT 000739|Hs16 241| Homo sapiens 16q22.1 sequence /len=124645*>ref|NT 000740|Hs16 242| Homo sapiens 16q22.2 sequence
/len=189134*>ref|NT 000741|Hs16 243| Homo sapiens 16q22 sequence /len=38542*>ref|NT 000742|Hs16 244| Homo sapiens 16q22.2 sequence
/len=180551*>ref|NT 000743|Hs16 245| Homo sapiens 16q24 sequence /len=109220*>ref|NT 001440|Hs16 1527| Homo sapiens 16p13.3 sequence
/len=252148*>ref|NT 001441|Hs16 1528| Homo sapiens 16q22.1 sequence /len=191565*>ref|NT 001442|Hs16 1529| Homo sapiens chr.16 sequence
/len=141313*>ref|NT 001480|Hs16 1612| Homo sapiens 16p13.3 sequence /len=253915*>ref|NT 001573|Hs16 1692| Homo sapiens 16p13.3 sequence
/len=167525*>ref|NT 001575|Hs16 1694| Homo sapiens 16p13.3 sequence /len=191145*>ref|NT 001576|Hs16 1695| Homo sapiens 16p13.3 sequence
/len=174098*>ref|NT 001577|Hs16 1696| Homo sapiens 16q21-q22 sequence /len=216021*>ref|NT 001578|Hs16 1697| Homo sapiens 16q21-q22 sequence
/len=157838*>ref|NT 001633|Hs16 1753| Homo sapiens chr.16 sequence /len=168731*>ref|NT 001634|Hs16 1754| Homo sapiens chr.16 sequence
/len=167080*>ref|NT 001699|Hs16 1820| Homo sapiens 16p13.3 sequence /len=161264*>ref|NT 001918|Hs16 2036| Homo sapiens 16p13.3 sequence
/len=31513*>ref|NT 001919|Hs16 2037| Homo sapiens 16p13.3 sequence /len=20733*>ref|NT 001936|Hs16 2054| Homo sapiens chr.16 sequence
/len=200594*>ref|NT 001971|Hs16 2089| Homo sapiens 16p13.3 sequence /len=54132*>ref|NT 002067|Hs16 2186| Homo sapiens chr.16 sequence
>ref|NT 002398|Hs16 2519| Homo sapiens 16p13.3 sequence /len=20612*/len=37476*>ref|NT 002112|Hs16 2231| Homo sapiens 16p13.3 sequence /len=25815

Thanks to:
JOEL MARTIN, CLIFF HAN, LAURIE A. GORDON, ASTRID TERRY, SHYAM PRABHAKAR, XINWEI SHE, GARY XIE, UFFE HELLSTEN,
YEE MAN CHAN, MICHAEL ALTHERR, OLIVIER COURONNE, ANDREA AERTS, EVA BAJOREK, STACEY BLACK,
HEATHER BLUMER, ELBERT BRANSCOMB, NANCY C. BROWN, WILLIAM J. BRUNO, JUDITH M. BUCKINGHAM,
DAVID F. CALLEN, CONNIE S. CAMPBELL, MARY L. CAMPBELL, EVELYN W. CAMPBELL, CHENIER CAOILE,
JEAN F. CHALLACOMBE, LESLIE A. CHASTEEN, OLGA CHERTKOV, HAN C. CHI, MARI CHRISTENSEN, LYNN M. CLARK,
JUDITH D. COHN, MIRIAN DENYS, JOHN C. DETTER, MARK DICKSON, MIRA DIMITRIJEVIC-BUSSOD, JULIO ESCOBAR,
JOSEPH J. FAWCETT, DAVE FLOWERS, DEA FOTOPULOS, TIJANA GLAVINA, MARIA GOMEZ, EIDELYN GONZALES,
DAVID GOODSTEIN, LYNNE A. GOODWIN, DEBORAH L. GRADY, IGOR GRIGORIEV, MATTHEW GROZA, NANCY HAMMON,
TREVOR HAWKINS, LAUREN HAYDU, CARL E. HILDEBRAND, WAYNE HUANG, SANJAY ISRANI, JAMIE JETT, PHILLIP B. JEWETT,
KRISTEN KADNER, HEATHER KIMBALL, ARTHUR KOBAYASHI, MARIE-CLAUDE KRAWCZYK, TINA LEYBA,
JONATHAN L. LONGMIRE, FREDERICK LOPEZ, YUNIAN LOU, STEVE LOWRY, THOM LUDEMAN, CHITRA F. MANOHAR,
GRAHAM A. MARK, KIMBERLY L. MCMURRAY, LINDA J. MEINCKE, JENNA MORGAN, ROBERT K. MOYZIS, MARK O. MUNDT,
A. CHRISTINE MUNK, RICHARD D. NANDKESHWAR, SAM PITLUCK, MARTIN POLLARD, PAUL PREDKI, BEVERLY PARSON-QUINTANA, LUCIA RAMIREZ, SAM RASH, JAMES RETTERER, DARRYL O. RICKE, DONNA L. ROBINSON, ALEX RODRIGUEZ,
ASAF SALAMOV, ELIZABETH H. SAUNDERS, DUNCAN SCOTT, TIMOTHY SHOUGH, RAYMOND L. STALLINGS,
MALINDA STALVEY, ROBERT D. SUTHERLAND, ROXANNE TAPIA, JUDITH G. TESMER, NINA THAYER, LINDA S. THOMPSON,
HOPE TICE, DAVID C. TORNEY, MARY TRAN-GYAMFI, MING TSAI, LEVY E. ULANOVSKY, ANNA USTASZEWSKA, NU VO,
P. SCOTT WHITE, ALBERT L. WILLIAMS, PATRICIA L. WILLS, JUNG-RUNG WU, KEVIN WU, JOAN YANG, PIETER DEJONG,
DAVID BRUCE, NORMAN A. DOGGETT, LARRY DEAVEN, JEREMY SCHMUTZ, JANE GRIMWOOD, PAUL RICHARDSON,
DANIEL S. ROKHSAR, EVAN E. EICHLER, PAUL GILNA, SUSAN M. LUCAS, RICHARD M. MYERS, EDWARD M. RUBIN &
LEN A. PENNACCHIO, authors of "The sequence and analysis of duplication-rich human chromosome 16"
NATURE, Vol. 432 p 988 and following. 23/30 December 2004.

Thirteen
million, nine
hundred
and twenty
six thousand
eight
hundred
and fifty five

Fourteen million one hundred and thirty-nine thousand three hundred and twenty-seven

Fourteen million three hundred and sixty one thousand six hundred and eighty two

fourteen
million six
hundred
and eighty
five
thousand
five
hundred
and seven

fourteen
million eight
hundred
and sixteen
thousand
and
seventy-five

eleven
million
thirty-three
thousand
one
hundred
and sixty
two

three billion two hundred and sixty thousand five hundred and seventy

fifteen
million four
hundred
and eighty
two
thousand
eight
hundred
and
seventy
seven

three
billion
seven
hundred
and ten
thousand six
hundred
and eleven

Fifteen
million nine
hundred
and thirty-
four
thousand
four
hundred
and thirty-
two

sixteen
million
nine
hundred
and twenty
three
thousand
eight
hundred
and sixty

Sixteen
million three
hundred
and
seventy
three
thousand
eight
hundred
and
nineteen

sixteen
million five
hundred
and forty
one
thousand

and nine
hundred

The Build 34 human genome assembly file for chromosomal sequence 16 is published by Project Gutenberg as Etext 11790. It was first released on 1 March 2004. It contains ninety million forty-one thousand nine hundred and thirty-two characters. In the *Nature* article (see page 525) the authors report that at time of writing there were still areas of missing sequence however seventy-eight million eight hundred and eighty-four thousand seven hundred and fifty-four base pairs of finished euchromatic sequence had been generated.

www.ingramcontent.com/pod-product-compliance
Lightning Source LLC
LaVergne TN
LVHW061217100826
845148LV00004B/780
9780955309229